AF556025

Refresher Guide in Veterinary Pharmacology and Toxicology

NIPA® GENX ELECTRONIC RESOURCES & SOLUTIONS P. LTD.
New Delhi-110 034

About the Editors

Dr. Amit Shukla, did his B.V.Sc. & A.H. and M.V.Sc. in Veterinary Pharmacology and Toxicology from G.B. Pant University of Agriculture and Technology, Pantnagar, Uttarakhand in the year 2010 and 2012, respectively. He completed his Ph.D. in Veterinary Pharmacology from Indian Veterinary Research Institute (IVRI), Izatnagar, Bareilly in the year 2015. He has more than 7 years of teaching experience in the field of Veterinary Pharmacology and Toxicology. He was awarded with several award viz. Best Educationist Award, CARS Project Award from DRDO-INMAS, Ministry of Defence, Government of India, New Delhi, Best poster Award and Best presentation Award etc. Presently, he is working as an Assistant Professor in the Department of Veterinary Pharmacology and Toxicology, College of Veterinary Science and Animal Husbandry, UP Pandit Deen Dayal Upadhyaya Pashu Chikitsa Vigyan Vishwavidyalaya Evam Go Anusandhan Sansthan (DUVASU), Mathura- 281001, Uttar Pradesh, India.

Prof. (Dr.) Atul Prakash did his B.V.Sc.& A.H. from Harbinger of Green Revolution G.B. Pant University of Agriculture and Technology, Pantnagar, Uttarakhand in the year 2004 and M.V.Sc. in Veterinary Pharmacology and Toxicology from N.D University of Agriculture and Technology, Ayodhya in the year 2006. Further he conferred with Doctor of Philosophy in Veterinary Pharmacology from Indian Veterinary Research Institute, Izatnagar, Bareilly in the year 2009. He has more than 15 years of experience in teaching of Pharmacology and Toxicology. He is awarded with several awards viz. Foreign Travel Grant from DBT, Govt. of India and CSIR, Govt. of India etc. He has guided more than 10 M.V.Sc. research scholars. Currently he is serving as Professor and Head in the Department of Veterinary Pharmacology and Toxicology, C.V.Sc. & A.H., U.P. Veterinary University (DUVASU), Mathura.

Dr. Soumen Choudhury did his B.V.Sc. & A.H. from CAU, Imphal in the year 2005 and M.V.Sc. in Veterinary Pharmacology and Toxicology from U.P. Veterinary University (DUVASU), Mathura in the year 2007. He completed his Ph.D. in Veterinary Pharmacology from Indian Veterinary Research Institute (IVRI), Izatnagar, Bareilly in the year 2014. He has more than 12 years of teaching experience in the field of Veterinary Pharmacology and Toxicology. He was awarded with several award viz. Prof. V.V. Ranade Young Scientist Award by Indian Society of Veterinary Pharmacology & Toxicology (ISVPT) etc. Presently, he is working as an Assistant Professor in the Department of Veterinary Pharmacology and Toxicology, College of Veterinary Science and Animal Husbandry, UP Pandit Deen Dayal Upadhyaya Pashu Chikitsa Vigyan Vishwavidyalaya Evam Go Anusandhan Sansthan (DUVASU), Mathura-281001, Uttar Pradesh, India.

Dr. Sakshi Tiwari did her B.V.Sc. & A.H. from NDVSU, Jabalpur in the year 2013 and M.V.Sc. in Veterinary Pathology from G.B. Pant University of Agriculture and Technology, Pantnagar, Uttarakhand in the year 2015. She completed her Ph.D. in Veterinary Pathology from LUVAS, Hisar in the year 2018. She worked as Veterinary Assistant Surgeon for five years in the State Government of M.P. She was awarded with several award viz. Dr. P.O. George Memorial Award 2019 for best Clinical Article from Indian Society of Veterinary Surgery, Best Young Lady Veterinarian Award etc. Presently, She is working as an Assistant Professor in the Department of Veterinary Pathology, College of Veterinary Science and Animal Husbandry, UP Pandit Deen Dayal Upadhyaya Pashu Chikitsa Vigyan Vishwavidyalaya Evam Go Anusandhan Sansthan (DUVASU), Mathura- 281001, Uttar Pradesh, India.

Refresher Guide in Veterinary Pharmacology & Toxicology

Amit Shukla
Assistant Professor
Department of Veterinary Pharmacology and Toxicology
College of Veterinary Science and Animal Husbandry
UP Pandit Deen Dayal Upadhyaya Pashu Chikitsa Vigyan
Vishwavidyalaya Evam Go Anusandhan Sansthan
(DUVASU), Mathura-281001, Uttar Pradesh, India

Atul Prakash
Professor and Head
Department of Veterinary Pharmacology and Toxicology
College of Veterinary Science and Animal Husbandry
UP Pandit Deen Dayal Upadhyaya Pashu Chikitsa Vigyan
Vishwavidyalaya Evam Go Anusandhan Sansthan
(DUVASU), Mathura-281001, Uttar Pradesh, India

Soumen Choudhury
Assistant Professor
Department of Veterinary Pharmacology and Toxicology
College of Veterinary Science and Animal Husbandry
UP Pandit Deen Dayal Upadhyaya Pashu Chikitsa Vigyan
Vishwavidyalaya Evam Go Anusandhan Sansthan
(DUVASU), Mathura-281001, Uttar Pradesh, India

Sakshi Tiwari
Assistant Professor
Department of Veterinary Pathology
College of Veterinary Science and Animal Husbandry
UP Pandit Deen Dayal Upadhyaya Pashu Chikitsa Vigyan
Vishwavidyalaya Evam Go Anusandhan Sansthan
(DUVASU), Mathura-281001, Uttar Pradesh, India

NIPA® GENX ELECTRONIC RESOURCES & SOLUTIONS P. LTD.
New Delhi-110 034

NIPA® GENX ELECTRONIC RESOURCES & SOLUTIONS P. LTD.

101,103, Vikas Surya Plaza, CU Block
L.S.C.Market, Pitam Pura, New Delhi-110 034
Ph : +91 11 27341616, 27341717, 27341718
E-mail: newindiapublishingagency@gmail.com
www: www.nipabooks.com

For customer assistance, please contact
Phone: + 91-11-27 34 17 17
Fax: + 91-11-27 34 16 16

Print ISBN: 978-93-58879-85-8

ebook ISBN: 978-93-58874-99-0

Composed and Designed by NIPA®.

Editorial Preface

Pharmacology and Toxicology is a subject that imposes a great responsibility on the veterinary and medical practitioner in their day to day clinical practice. Conceptualizing the concepts and making them at ease to the budding vets is the need of the hour. The central dogma of revisiting the concepts lies well with the practices of the objective type questions to help individuals in attaining a great success in their academic and professional pursuit. This book will provide a comprehensive detailing of the subject and build their analytical approach to mitigate the academic challenges in the competitive exams viz. ICAR-SRF, JRF, NET, ARS, CSIR-NET, UGC-NET, NIPER, IAS, IFS, PCS etc.

This book contains 11 chapters encompassing the multiple choice questions, true false and fill in the blanks vis a vis matching to consolidate the scientific acumen of the students in brightening their future. The main highlight of this book is the vast coverage of new chapters related to drug discovery, multi drug resistance (MDR) and clinical pharmacology and Veterinary therapeutics. Additionally book also provides a practice question sets to the students. All the chapters are covered as per the new academic regulations of VCI and ICAR - Broad Subject Matter Area (BSMA).

We wish all the success to the readers for their future academic career.

Editors

Contents

1

Source and Nature of Drugs and Pharmacokinetics

Amit Shukla, Soumen Choudhury and Atul Prakash

Q1. The exit of drug from C.S.F. irrespective of its entry is by

a) Passive diffusion
b) Filtration
c) Facilitated diffusion
d) All of the above

Q2. The number of half-lives required to reach 99% of steady state concentration is

a) 3.3
b) 4.4
c) 5.5
d) 6.6

Q3. The receptor concept was first introduced by

a) J.N. Langley in 1878
b) Simonis in 1964
c) Paul Ehrlich in 1926
d) Walksman in 1826

Q4. Characteristic untoward symptoms unrelated to dose or pharmacodynamic property of the drug is called

a) Idiosyncrasy
b) Toxic effect
c) Hypersensitivity
d) Intolerance

Q5. Identify the incorrect matching out of the following synthetic reactions and corresponding endogenous substrates:

a) Glucoronidation: UDPGA
b) Sulphate conjugation: PAP
c) Methylation: SAM
d) Amino acid Conjugation: GSH

Q6. A low extent sulphate conjugation of arylamines occurs in

a) Dogs
b) Equines
c) Pigs
d) Ovine

Q7. Which of the following is not required for oxidative reaction by MFOs

a) Cyt P_{450}
b) Oxygen
c) H_2O
d) Reduced NADP

Q8. Which of the following deals with the quantitative measurements of pharmacological effects of a drug in the body

a) Pharmacokinetics
b) Metrology
c) Posology
d) Pharmacometrics

Q9. Metabolism by N-acelytation occurs in xenobiotics containing--- groups

a) R-NH2
b) R-NH2 and R-NH-NH2
c) R-SH group
d) R-OH group

Q10. Co-factor required for methylation of xenobiotics is

a) Acetyl cysteine
b) S-adenosyl methionine
c) 3' Phospho adenosine 5' phosphosulphate
d) Glutathione

Q11. Which of the following transporters are responsible for secretion of organic cationic drugs proximal renal tubules?

a) MDR1
b) ATP-binding cassette
c) MRP2
d) P- glycoprotein

Q12. How many molecule(s) of oxygen is/are required for oxidation of a substrate by a Cytochrome P_{450}

a) Two
b) Four
c) One
d) Three

Q13. Which of the following CYP isomers has maximum share in phase I metabolism of drugs?

a) CYP2C8/9
b) CYP2B6
c) CYP3A4
d) CYP2D6

Q14. Who compiled first pharmacopoeia ?

a) Theophrastus
b) Valerius Cordus
c) MJB Orifila
d) Christopher Wren

Q15. The rate theory of drug action was introduced by

a) Clark
b) Paton
c) Stephenson
d) Katz

Q16. Enzyme that catalyses the hydration of CO2 in renal tubules

a) Carbonic anhydrase
b) Glucuronyl transferase
c) Glutathione –s- transferase
d) Na-K ATPase

Q17. If at a moment the total amount of a drug in the body is 10 g and plasma concentration is 125 µg/ml, the Vd of the drug will be

a) 40 L
b) 16 L
c) 80 L
d) 8 L

Q18. How many membrane bound isoforms of adenylyl cyclase (AC) are found in mammalian tissues ?

a) 5
b) 7
c) 8
d) 9

Q19. Pharmacokinetics is:

a) The study of absorption, distribution, metabolism and excretion of drugs
b) The study of biological and therapeutic effects of drugs
c) The study of mechanisms of drug action
d) The study of methods of new drug development

Q20. What does "pharmacokinetics" include?

a) Influence of drugs on genes
b) Influence of drugs on metabolism processes
c) Drug biotransformation in the organism
d) Complications of drug therapy

Q21. The main mechanism of most drugs absorption in GI tract is:

a) Endocytosis and exocytosis
b) Passive diffusion (lipid diffusion)
c) Filtration (aqueous diffusion)
d) Active transport (carrier-mediated diffusion)

Q22. What kind of substances can't permeate membranes by passive diffusion?

a) Hydrophobic substances
b) Non-ionized substances
c) Lipid-soluble
d) Hydrophilic substances

Q23. What is true regarding active transport of drugs?

a) Engulf of drug by a cell membrane with a new vesicle formation
b) Transport against concentration gradient
c) Transport without energy consumption
d) Transport of drugs trough a membrane by means of diffusion

Q24. Pick out the appropriate alimentary route of administration when passage of drugs through liver is minimized:

a) Intraduodenal
b) Rectal
c) Transdermal
d) Oral

Q25. Tick the feature of the sublingual route:

a) A drug can be administrated in a variety of doses
b) A drug is exposed more prominent liver metabolism

c) A drug is exposed to gastric secretion

d) Very rapid absorption

Q26. What is characteristic of the intramuscular route of drug administration?

a) Only water solutions can be injected

b) The action develops slower, than at oral administration

c) Oily solutions can be injected

d) Opportunity of hypertonic solution injections

Q27. Most of drugs are distributed homogeneously.

a) True
b) False

c) statement is incomplete
d) both a and c

Q28. Biological barriers include all except

a) Renal tubules
b) Cell membranes

c) Placenta
d) Capillary walls

Q29. What is the reason of complicated penetration of some drugs through brain-blood barrier?

a) Meningitis

b) High lipid solubility of a drug

c) Absence of pores in the brain capillary endothelium

d) High endocytosis degree in a brain capillary

Q30. For the calculation of the volume of distribution (Vd) one must take into account:

a) A daily dose of drug

b) Therapeutical width of drug action

c) Concentration of substance in urine

d) Concentration of a substance in plasma

Q31. The term "biotransformation" includes the following:

a) Accumulation of substances in a tissue

b) Process of physicochemical and biochemical alteration of a drug in the body

c) Binding of substances with plasma proteins

d) Accumulation of substances in a fat tissue

Q32. Tick the drug type for which microsomal oxidation is the most prominent:

a) Water soluble
b) High molecular weight

c) Low molecular weight
d) Lipid soluble

Q33. Elimination rate constant (Kelim) is defined by the following parameter:

a) Highest single dose
b) Half life (t ½)
c) Maximal concentration of a substance in plasma
d) Rate of absorption

Q34. Systemic clearance (CLs) is related with:

a) Bioavailability and half life
b) Only the elimination rate constant
c) Only the concentration of substances in plasma
d) Volume of distribution, half life and elimination rate constant

Q35. Elimination is expressed as follows:

a) Time required to decrease the amount of drug in plasma by one-half
b) Clearance speed of some volume of blood from substance
c) Rate of renal tubular reabsorption
d) Clearance of an organism from a xenobiotic

Q36. Elimination includes

a) Excretion only
b) Metabolism only
c) Metabolism and excretion both
d) None of the above

Q37. Half life (t ½) doesn't depend on:

a) Time of drug absorption
b) Biotransformation
c) Concentration of a drug in plasma
d) Rate of drug elimination

Q38. In case of liver disorders accompanied by a decline in microsomal enzyme activity the duration of action of some drugs is:

a) Remained unchanged
b) Increased
c) Decreased
d) Changed insignificantly

Q39. Conjugation of a drug includes the following EXCEPT:

a) Methylation
b) Sulfate formation
c) Glucoronidation
d) Oxidation

Q40. Conjugation is:

a) Process of drug oxidation by special oxidases
b) Coupling of a drug with an endogenous substrate
c) Process of drug reduction by special enzymes
d) Solubilization in lipids

Q41. Stimulation of liver microsomal enzymes can:

a) Require the dose decrease of some drugs
b) Require the dose increase of some drugs

c) Intensify the unwanted reaction of a drug

d) Prolong the duration of the action of a drug

Q42. What does "affinity" mean?

a) A measure of bioavailability of a drug

b) A measure of inhibiting potency of a drug

c) A measure of how tightly a drug binds to a receptor

d) A measure of how tightly a drug binds to plasma proteins

Q43. Target proteins which a drug molecule binds are:

a) Only carriers
b) Only receptors
c) Only ion channels
d) All of the above

Q44. Irreversible interaction of an antagonist with a receptor is due to:

a) Hydrogen bonds
b) Ionic bonds
c) Covalent bonds
d) All of the above

Q45. Mechanisms of transmembrane signaling are the following EXCEPT:

a) Ligand-gated ion channels that can be induced to open or close by binding a ligand

b) Trans membrane receptors that bind and stimulate a protein tyrosine kinase

c) Trans membrane receptor protein that stimulates a GTP-binding signal transducer protein (G-protein) which in turn

d) Gene replacement by the introduction of a therapeutic gene to correct a genetic effect

Q46. Tick the second messenger of metabotropic receptor:

a) Sodium ions
b) cAMP
c) Phospholipase C
d) Adenylyl cyclase

Q47. Tick the substance which changes the activity of an effector element but doesn't belong to second messengers:

a) G–protein
b) cGMP
c) cAMP
d) Calcium ions

Q48. The increase of second messengers' (cAMP, cGMP, Ca2+ etc.) concentration leads to:

a) Antagonism with endogenous ligands

b) Blocking of interaction between a receptor and an effector

c) Inhibition of intracellular protein kinases and protein phosphorylation

d) Proteinkinases activation and protein phosphorylation

Q49. Fastest action is produced through

a) GPCR
b) LGIC
c) Enzyme linked receptors
d) Nuclear receptors

Q50. Zinc fingers are present in

a) GPCR
b) LGIC
c) Enzyme linked receptors
d) Nuclear receptors

Q51. Serpentine receptors is also known as

a) GPCR
b) LGIC
c) Enzyme linked receptors
d) Nuclear receptors

Q51. Petals of lilly is characteristic of

a) GPCR
b) LGIC
c) Enzyme linked receptors
d) Nuclear receptors

Q52. Serpentine receptors is also known as

a) GPCR
b) LGIC
c) Enzyme linked receptors
d) Nuclear receptors

Q53. Insulin acts through

a) GPCR
b) LGIC
c) Enzyme linked receptors
d) Nuclear receptors

Q54. Steroid receptors are also known as

a) Transcription factors
b) Nuclear receptor
c) Both of the above
d) None of the above

Q55. Vonslyke equation is related with

a) Metabolism
b) Clearance
c) Volume of distribution
d) Diffusion

Q56. Ficks law governs

a) Rate of metabolism
b) Rate of diffusion
c) Both of the above
d) Clearance

Q57. Microsomal enzymes are not required for

a) Glucoronidation
b) Sulphation
c) Oxidation
d) Reduction

Q58. All of the following statements about efficacy and potency are true EXCEPT:

a) Potency is a comparative measure, refers to the different doses of two drugs that are needed to produce the sameeffect

b) Efficacy is the maximum effect of a drug

c) The ED50 is a measure of drug's efficacy

d) Efficacy is usually a more important clinical consideration than potency

Q59. Which effect may lead to toxic reactions when a drug is taken continuously or repeatedly?

a) Tolerance
b) Tachyphylaxis
c) Cumulative effect
d) Refractoriness

Q60. Enzyme responsible for are responsible for the hydrolysis of the cyclic 3, 5-phosphodiester bond found in cyclic AMP and cyclic GMP

a) Tyrosine kinase
b) Phospholipase
c) Protein kinase
d) Phosphodiesterase (PDE)

Q61. Sildenafil produces vasodilation through inhibition of

a) Tyrosine kinase
b) hospholipase
c) Protein kinase
d) Phosphodiesterase (PDE)

Q62. Name the enzyme involved in maintenance of ionic gradients across the cell membrane.

a) Calcium ATPase
b) Na+, K+-ATPase
c) Mg ATPase
d) All of the above

Q63. Histamine acts by H1 receptor-

a) Gq-PLC- IP3
b) Gs-cAMP
c) Both of the above
d) Through receptor tyrosine kinase pathway

Q64. What term is used to describe a more gradual decrease in responsiveness to a drug, taking days or weeks to develop?

a) Tachyphylaxis
b) Cumulative effect
c) Refractoriness
d) Tolerance

Q65. What term is used to describe a decrease in responsiveness to a drug which develops in a few minutes?

a) Tolerance
b) Tachyphylaxis
c) Refractoriness
d) Cumulative effect

Q66. Father of Indian pharmacology

a) R N CHOPRA
b) SUSRUTA
c) CHARAK
d) VAGBHATTA

Q67. Tachyphylaxis is:

a) A decrease in responsiveness to a drug, taking days or weeks to develop

b) Very rapidly developing tolerance

c) A drug interaction between two similar types of drugs

d) None of the above

Q68. Tolerance and drug resistance can be a consequence of:

a) Increased receptor sensitivity

b) Decreased metabolic degradation

c) Decreased renal tubular secretion

d) Change in receptors, loss of them or exhaustion of mediators

Q69. The situation when failure to continue administering the drug results in serious psychological and somatic disturbances iscalled?

a) Idiosyncrasy
b) Refractoriness

c) Tachyphylaxis
d) Abstinence syndrome

Q70. What is the type of drug-to-drug interaction which is the result of interaction at receptor, cell, enzyme or organ level?

a) Pharmacokinetic interaction

b) Pharmaceutical interaction

c) Pharmacodynamic interaction

d) Physical and chemical interaction

Q71. What is the type of drug-to-drug interaction which is connected with processes of absorption, biotransformation, distribution and excretion?

a) Pharmaceutical interaction

b) Pharmacokinetic interaction

c) Physical and chemical interaction

d) Pharmacodynamic interaction

Q72. What does the term "potentiation" mean?

a) Fast tolerance developing

b) Intensive increase of drug effects due to their combination

c) Hypersensitivity to a drug

d) Cumulative ability of a drug

Q73. The term "chemical antagonism" means that:

a) Two drugs combine with one another to form a more active compound

b) Two drugs combine with one another to form a more water soluble compound

c) Two drugs combine with one another to form a more fat soluble compound

d) Two drugs combine with one another to form an inactive compound

Q74. A teratogenic action is:

a) Toxic action on blood system

b) Toxic action on kidneys

c) Negative action on the fetus causing fetal malformation

d) Toxic action on the liver

Q75. Characteristic unwanted reaction which isn't related to a dose or to a pharmacodynamic property of a drug is called:

a) Tolerance

b) Teratogenic action

c) Idiosyncrasy

d) Hypersensitivity

Q76. Therapeutic index (TI) is:

a) A ratio used to evaluate the effectiveness of a drug

b) A ratio used to evaluate the bioavailability of a drug

c) A ratio used to evaluate the elimination of a drug

d) A ratio used to evaluate the safety and usefulness of a drug for indication

Q77. Idiosyncratic reaction of a drug is:

a) Quantitatively exaggerated response

b) Unpredictable, inherent, qualitatively abnormal reaction to a drug

c) A type of drug antagonism

d) A type of hypersensitivity reaction

Q78. If an agonist can produce submaximal effects and has moderate efficacy it's called:

a) Full agonist

b) Agonist-antagonist

c) Partial agonist

d) Antagonist

Q79. Aquaporins can be blocked by

a) Carbon di oxide

b) para- chloromercurobenzene sulfonate

c) Nitrogen di oxide

d) None of the above

Q80. Passive transfer of the drug is also known as

a) Up hill transport

b) Down hill transport

c) Facilitated diffusion

d) None of the above

Q81. The drug molecules could easily pass via filtration if they have molecular weight

a) More than 1000 dalton
b) Less than 100 dalton
c) 100-500 dalton
d) 500-1000 dalton

Q82. Basic drugs are transported across the membrane via

a) Organic anion transporters
b) Organic cation transporters
c) P glycoproteins
d) Filtration

Q83. Agonist should possess:

a) Affinity and efficacy
b) Only efficacy
c) Potency but no affinity
d) Affinity but no efficacy

Q84. Penicillin and other acidic drugs can cross biological membrane with the help of

a) Organic anion transporters
b) Organic cation transporters
c) P glycoprotein
d) Filtration

Q85. Example of efflux transporter is

a) MDR
b) P glycoproteins
c) Both of the above
d) None of the above

Q86. SLC6A3- is helpful in transportation of FOR

a) Nor epinephrine
b) Seratonin
c) Histamine
d) DOPAMINE

Q87. SLC6A4 is helpful in transportation of FOR

a) Nor epinephrine
b) Histamine
c) SERATONIN
d) DOPAMINE

Q88. SLC6A5 is helpful in transportation of FOR

a) Nor epinephrine
b) Histamine
c) SERATONIN
d) DOPAMINE

Q89. Absorption of vitamin B1, B2 and B12 occurs through

a) Passive transport
b) Facilitated diffusion
c) Active transport
d) Pinocytosis

Q90. Active transport is also known as

a) Up hill transport
b) Down hill transport
c) Facilitated diffusion
d) None of the above

Q91. P glycoprotein was discovered by

a) Paton
b) Juliano R.L. and Ling
c) Clark
d) Katz

Q92. Toxicity of ivermectin in collie breed is due to lack of

a) MDR2
b) MDR1
c) MDR3
d) MDR5

Q93. Botulinum and diphtheria toxin are transported via

a) Endocytosis
b) Facilitated diffusion
c) Reverse transport
d) Down hill transport

Q94. Unit of area under curve is

a) Mg per hour
b) mg/ml hour
c) ml hour
d) Ml per hour

Q95. Bioavailability of a drug through Intravenous route is

a) 20%
b) 60%
c) 100%
d) Zero

Q96. Wider distribution or extensive tissue binding of drugs will lead to

a) Increase in volume of distribution
b) Decrease in volume of distribution
c) No change in Vd
d) None of the above

Q97. Tie to reach 95% of steady state concentration is achieved in

a) 3.3 t1/2
b) 4.3 t1/2
c) 5.5 t1/2
d) 6.6 t1/2

Q98. Physiological model of kinetics is also known as

a) Compartment model
b) Non compartment model
c) Perfusion rate limited model
d) Statistical model

Q99. Fixed fraction of the drug is processed per unit time in

a) Zero order kinetics
b) Second order kinetics
c) First order kinetics
d) Saturation kinetics

Q100. Fixed fraction of the drug is processed per unit time in

a) Zero order kinetics
b) Second order kinetics
c) First order kinetics
d) Saturation kinetics

Q101. Half liferemain constant in

a) Zero order kinetics
b) Second order kinetics
c) First order kinetics
d) Saturation kinetics

Q102. Therapeutic ratio is calculated by

a) LD50/ED50
b) LD25/ED55
c) LD25/ED75
d) LD75/ED25

Q103. Standard safety margin is calculated by

a) LD99/ED1
b) LD50/ED99
c) LD1/ED99
d) LD75/ED90

Q104. Following are the examples of naturally obtained drug except

a) Lobeline
b) Nicotine
c) Heparin
d) Sulphonamide

Q105. Gingerol is obtained from .... part of ginger

a) Root
b) Stem
c) Leaves
d) Rhizome

Q106. Antitetanus serum belongs to which source

a) Plant
b) Mineral
c) Animal
d) None of these

Q107. Strychinine is obtained from

a) Kaner
b) Nux vomica plant
c) Rouwolfia serpentina
d) Tobacco plant

Q108. Buprenorphine is obtained by .... nucleus

a) Morphine
b) Thebaine
c) Cocaine
d) Dimethyl morphine

Q109. Recombinant DNA technology produces

a) Insulin
b) Bovine somatotrophine
c) Apomorphine
d) Muscarine

Q110. Alkaloid in liquid state do not posses

a) Carbon
b) Hydrogen
c) Oxygen
d) Nitrogen

Q111. Alkaloid in solid state do posses

a) Carbon
b) Hydrogen
c) Oxygen
d) All of the above

Q112. Alkaloid can be synthesized by utilizing

a) Purine ring
b) Pyridine ring
c) Pyrimidine ring
d) Acrylic ring

Q113. Example of liquid alkaloid is

a) Morphine
b) Atropine
c) Arecoline
d) Quinine

Q114. Which test is used to detect alkaloid

a) Kedde test
b) Dragendroff test
c) Foam test
d) Legal test

Q115. Alkaloid can be oxidized by

a) Tannic acid
b) KMnO4
c) Picric acid
d) All of the above

Q116. Sugar moiety linked with nonsugar moiety with ether linkage in

a) Alkaloid
b) Glycosides
c) Proteins
d) Tannin

Q117. Example of cyanogenetic glycoside is

a) Senna
b) Digoxin
c) Digitoxin
d) Dhurrin

Q118. Foam test is done to detect

a) Alkaloid
b) Saponin
c) Proteins
d) Tannin

Q119. Glycyrrhizin is an example of

a) Alkaloid
b) Proteins
c) Saponin
d) Tannin

Q120. Legal test is done to detect

a) Anthraquinone glycosides
b) Cardiac glycosides
c) Cyanogentic glycosides
d) Saponin

Q121. Borntrager test is done to detect

a) Anthraquinone glycosides
b) Cardiac glycosides
c) Cyanogentic glycosides
d) Saponin

Q122. Guinard test is done to detect

a) Anthraquinone glycosides
b) Cardiac glycosides
c) Cyanogentic glycosides
d) Saponin

Q123. Picrate paper test is done to detect

a) Anthraquinone glycosides
b) Cardiac glycosides
c) Cyanogentic glycosides
d) Saponin

Q124. Ethereal oil is synonymous to

a) Volatile oil
b) Fixed oil
c) Mineral oil
d) All of the above

Q125. Flavoring oil is also known as

a) Ethereal oil
b) Aromatic oil
c) Essential oil
d) All of the above

Q126. Alcoholic solutions of volatile oils is known as

a) Empyreumatic oil
b) Essences
c) Both of the above
d) None of the above

Q127. Thymol oil is in nature

a) Liquid
b) Soild
c) Watery
d) All of the above

Q128. Oils obtained after destructive distillation of the plant are known as

a) Empyreumatic oil
b) Essences
c) Both of the above
d) Saponin

Q129. Fixed oil are obtained by the process of.........

a) Distillation
b) Expression
c) Low pressure distillation
d) All of the above

Q130. Combination of higher alcohol with higher fatty acid is known as

a) Oil
b) Wax
c) Resin
d) Gum

Q131. Crude turpentine is an example of

a) Gum resin
b) Balsum
c) Oleo resin
d) Saponin

Q132. Mineral oil is obtained by

a) Fractional distillation
b) Destructive distillation
c) Expression
d) None of these

Q133. Complex phenolic compound having astringent property is defined as

a) Saponin
b) Resin
c) Tannin
d) Gum

Q134. Pyrogallol is derived from

a) Dihydric catechol
b) Trihydric phenol
c) Both of these
d) Oleoresins

Q135. Oak gall and chest nut are the examples of

a) Dihydric catechol
b) Trihydric phenol /pyrogallol
c) Pyrocatechol
d) Resins

Q136. Gold beater skin test is done to detect

a) Saponin b) Alkaloid
c) Resin d) Tannin

Q137. Ferric chloride precipitation test is done to detect

a) Saponin b) Alkaloid
c) Resin d) Tannin

Q138. Secretion of plants act as emulsifying agent and suspending agent for insoluble substances are known as

a) Alkaloids b) Resins
c) Gums d) Balsum

Q139. Tragacanth and agar are the examples of

a) Alkaloids b) Resins
c) Gums d) Balsum

Q140. Tartaric acid is found in

a) Tamarind b) Banana
c) Mango d) Orange

Q141. Nin hydrintet is done to detect

a) Protein b) Aminoacid
c) Lipids d) Alcohol

Q142. Omeprazole is mainly metabolized by

a) CYP3A4 b) CYP2C19
c) CYP2C9 d) CYP2D6

Q143. Rate limiting enzyme of porphyrin synthesis delta ALA synthase is induced by

a) Valproic acid b) Ketoconazole
c) Cimetidine d) Rifampicin

Q144. Example of enzyme inhibitor is all except

a) Valproic acid b) Ketoconazole
c) Cimetidine d) Rifampicin

Q145. In CYP3A4 , 3 DESIGNATES

a) Gene number b) Subfamily
c) Family d) Palindromic sequence

Q146. In CYP3A4 , A DESIGNATES

a) Gene number b) Subfamily
c) Family d) Palindromic sequence

Q147. In CYP3A4 , 4 DESIGNATES

a) Gene number
b) Subfamily
c) Family
d) Palindromic sequence

Q148. Vd is the determinant of

a) Maintenance dose
b) Loading dose
c) Clearance
d) All of the above

Q149. Examples of drugs showing zero order kinetics except

a) Warfarin
b) Aspirin
c) Theophylline
d) Ketoprofen

Q150. Dosage regimen includes

a) Frequency of dosing
b) Dose
c) Route of administration
d) All of the above

Q151. Essential drugs' are:

a) Life saving drugs
b) Drugs that must be present in the emergency bag of a doctor
c) Drugs that meet the priority health care needs of the population
d) Drugs that are listed in the pharmacopoia of a country

Q152. An 'orphan drug' is:

a) A drug which acts on Orphanin receptors
b) A very cheap drug
c) A drug needed for treatment or prevention of a rare disease
d) A drug which has no therapeutic use

Q153. Orphan drug act came in the year

a) 1987
b) 1986
c) 1983
d) 1985

Q154. Term pharmacognosy was given by

a) Stephen Defelice
b) Seydler
c) Stephenson
d) Paton

Q155. Nutraceutical term was coined by

a) Stephen Defelice
b) Seydler
c) Stephenson
d) Paton

Q156. The centre for National Pharmacovigilance Programme of India is located in

a) Lucknow
b) Delhi
c) Ghaziabad
d) Mumbai

Q157. Example of a chronobiotics is

a) Histamine
b) Melatonin
c) Adrenaline
d) Procaine

Q158. Branch of pharmacology deals with study of weight and measurement is known as

a) Pharmacometrics
b) Metrology
c) Posology
d) Pharmacognosy

Q159. Study of quantitative measurement of drug action and effect is known as

a) Pharmacometrics
b) Metrology
c) Posology
d) Pharmacognosy

Q160. First material medica was complied by

a) Valerius cordus
b) Wren
c) Alxender wood
d) Dioscoroides

Q161. Indian pharmacopoeia was launched in

a) 1950
b) 1951
c) 1955
d) 1953

Q162. Tailor made drug concept belongs to

a) Pharmacometrics
b) Pharmacogenomics
c) Pharmacogenetics
d) Pharmacoepidemiology

Q163. Drug administered through the following route is most likely to be subjected to first-pass metabolism:

a) Subcutaneous
b) Sublingual
c) Oral
d) Rectal

Q164. Which of the following drugs is administered by intranasal spray/application for systemic action:

a) Azelastine
b) Phenylephrine
c) Beclomethasone dipropionate
d) Desmopressin

Q165. Alkalinization of urine hastens the excretion of:

a) Strong electrolyte
b) Weakly basic drugs
c) Weakly acidic drugs
d) Nonpolar drugs

Q166. Which of the following is a weakly acidic drug:

a) Chloroquine phosphate
b) Atropine sulfate
c) Phenytoin sodium
d) Ephedrine hydrochloride

Q167. Bioavailability differences among oral formulations of a drug are most likely to occur if the drug:

a) Undergoes little first-pass metabolism
b) Is completely absorbed
c) Is freely water soluble
d) Is incompletely absorbed

Q168. The most important factor governing absorption of a drug from intact skin is:

a) Nature of the base used in the formulation
b) Lipid solubility of thedrug
c) Site of application
d) Molecular weight of the drug

Q169. Marked redistribution is a feature of:

a) Highly plasma protein bound drugs
b) Poorly lipid soluble drugs
c) Highly lipid soluble drugs
d) Depot preparations

Q170. The blood-brain barrier, which restricts entry of many drugs into brain, is constituted by:

a) Tight junctions between endothelial cells of brain capillaries
b) P-glycoprotein efflux carriers in brain capillary cells
c) Enzymes present in brain capillary walls
d) All of the above

Q171. High plasma protein binding:

a) Facilitates glomerular filtration of the drug
b) Increases volume of distribution of the drug
c) Generally makes the drug long acting
d) Minimises drug interactions

Q172. Which of the following is a prodrug:

a) Hydralazine
b) Captopril
c) Clonidine
d) Enalapril

Q173. The most commonly occurring conjugation reaction for drugs and their metabolites is:

a) Methylation
b) Acetylation
c) Glucuronidation
d) Glutathione conjugation

Q174. Microsomal enzyme induction can be a cause of:

a) Psychological dependence
b) Idiosyncrasy
c) Physical dependence
d) Tolerance

Q175. Which drug metabolizing reaction is entirely nonmicrosomal:

a) Reduction
b) Glucuronide conjugation
c) Oxidation
d) Acetylation

Q176. Select the drug that undergoes extensive first-pass metabolism in the liver:

a) Phenylbutazone
b) Propranolol
c) Phenobarbitone
d) Theophylline

Q177. The plasma half life of penicillin-G is longer in the new born because their:

a) Drug metabolizing enzymes are immature
b) Plasma protein level is low
c) Glomerular filtration rate is low
d) Tubular transport mechanisms are not well developed

Q178. If a drug has a constant bioavailability and first order elimination, its maintenance dose rate will be directly proportional to its:

a) Total body clearance
b) Plasma protein binding
c) Volume of distribution
d) Lipid solubility

Q179. The loading dose of a drug is governed by its:

a) Plasma half life
b) Volume of distribution
c) Renal clearance
d) Elimination rate constant

Q180. When the same dose of a drug is repeated at half life intervals, the steady-state (plateau) plasma drug concentration is reached after:

a) 4–5 half lives
b) 2–3 half lives
c) 6–7 half lives
d) 8–10 half lives

Q181. Monitoring plasma drug concentration is useful while using:

a) Lithium carbonate
b) MAO inhibitors
c) Levodopa
d) Antihypertensive drugs

Q182. Which of the following drugs acts by inhibiting an enzyme in the body:

a) Metoclopramide
b) Levodopa
c) Allopurinol
d) Atropine

Q183. The following is a competitive type of enzyme inhibitor:

a) Physostigmine
b) Theophylline
c) Disulfiram
d) Acetazolamide

Q184. Drugs acting through receptors exhibit the following features except:

a) Competitive antagonism
b) Dependence of action on lipophilicity
c) High potency
d) Structural specificity

Q185. Agonists affect the receptor molecule in the following manner:

a) Denature the receptor protein
b) Alter its amino acid sequence
c) Induce covalent bond formation
d) Alter its folding or alignment of subunits

Q186. Which of the following is a G protein coupled receptor:

a) Nicotinic cholinergic receptor
b) Insulin receptor
c) Muscarinic cholinergic receptor
d) Glucocorticoid receptor

Q187. The following receptor has an intrinsic ion channel:

a) Histamine H2 receptor
b) Histamine H1 receptor
c) GABA-benzodiazepine receptor
d) Adrenergic alfa receptor

Q188. Select the receptor that is located intracellularly:

a) Prostaglandin receptor
b) Angiotensin receptor
c) Opioid μ receptor
d) Steroid receptor

Q189. The receptor transduction mechanism with the fastest time-course of response effectuation is:

a) Adenyl cyclase-cAMP
b) AMP pathway
c) Phospholipase C-IP3-DAG Modulation of Protein Synthesis
d) Intrinsic ion channel operation

Q190. Down regulation of receptors can occur as a consequence of:

a) Continuous use of antagonists
b) Continuous use of agonists
c) Denervation
d) Chronic use of CNS depressants

Q191. Which of the following drugs exhibits 'therapeutic window' phenomenon:

a) Diazepam
b) Furosemide
c) Imipramine
d) Captopril

Q192. Which of the following is always true:

a) A more potent drug is safer

b) A more potent drug can produce the same response at lower doses

c) A more potent drug is clinically superior

d) A more potent drug is more efficacious

Q193. The therapeutic index of a drug is a measure of its:

a) Dose variability
b) Potency
c) Safety
d) Efficacy

Q194. If the effect of combination of two drugs is equal to the sum of their individual effects, the two drugs are exhibiting:

a) Potentiation
b) Cross tolerance
c) Antagonism
d) Additive effect

Q195. The antidotal action of sodium nitrite in cyanide poisoning is based on:

a) Physiological antagonism
b) Noncompetitive antagonism
c) Chemical antagonism
d) Physical antagonism

Q196. A drug which is generally administered in standard doses without the need for dose individualization is:

a) Prednisolone
b) Digoxin
c) Mebendazole
d) Insulin

Q197. Etomidate cleared by

a) One compartment model
b) Two compartment model
c) Three compartment model
d) All of the above

Q198. Mean residence time is calculated by

a) AUC/AUMC
b) AUMC/AUC
c) AUCIV/AUC ORAL
d) None of the above

Q199. Mayer test is done to detect

a) Alkaloid
b) Saponin
c) Glycosides
d) Resin

Q200. in is the suffix attached with

a) Alkaloid
b) Tannin
c) Glycosides
d) Resin

Answer Keys

1	c	2	d	3	a	4	a	5	d	6	c	7	c
8	d	9	b	10	b	11	b	12	c	13	c	14	b
15	b	16	a	17	c	18	d	19	a	20	c	21	b
22	d	23	b	24	b	25	d	26	c	27	b	28	a
29	c	30	d	31	b	32	d	33	b	34	d	35	d
36	c	37	a	38	b	39	d	40	b	41	b	42	c
43	d	44	c	45	d	46	b	47	a	48	d	49	b
50	d	51	b	52	a	53	c	54	c	55	b	56	b
57	b	58	c	59	c	60	d	61	d	62	b	63	a
64	d	65	b	66	a	67	b	68	d	69	d	70	c
71	b	72	d	73	d	74	c	75	d	76	d	77	d
78	c	79	b	80	b	81	b	82	b	83	a	84	a
85	c	86	d	87	c	88	a	89	b	90	a	91	b
92	b	93	a	94	b	95	c	96	a	97	b	98	c
99	c	100	a	101	c	102	c	103	c	104	d	105	d
106	c	107	b	108	b	109	b	110	c	111	d	112	b
113	c	114	b	115	b	116	b	117	d	118	b	119	c
120	d	121	a	122	c	123	c	124	a	125	d	126	b
127	b	128	a	129	b	130	b	131	c	132	a	133	c
134	b	135	b	136	d	137	d	138	c	139	c	140	a
141	b	142	b	143	d	144	d	145	c	146	b	147	a
148	b	149	d	150	d	151	c	152	c	153	c	154	b
155	a	156	c	157	b	158	b	159	a	160	d	161	c
162	b	163	c	164	d	165	c	166	c	167	d	168	b
169	c	170	d	171	c	172	d	173	c	174	d	175	d
176	b	177	d	178	a	179	b	180	a	181	a	182	c
183	a	184	b	185	d	186	c	187	c	188	d	189	d
190	b	191	c	192	b	193	c	194	d	195	c	196	c
197	c	198	b	199	a	200	c						

2

Mechanism and Problem of Drug Resistance

Amit Shukla, Soumen Choudhury and Atul Prakash

Q1. Resistance in tetracycline is due to mutational changes in

a) Tet A gene
b) Kat G
c) Par C par E
d) Inh A

Q2. Single step mutation can be seen in

a) Erythromycin
b) Tetracycline
c) Chloramphenicol
d) Rifampicin

Q3. Multistep mutation is observed in

a) Streptomycin
b) Rifampicin
c) Both A and C
d) Tetracycline

Q4. Transfer of resistance gene through environment free naked DNA is in

a) Conjugation
b) Transformation
c) Transduction
d) All of the above

Q5. Physical contact between bacteria responsible for drug resistance is

a) Conjugation
b) Transformation
c) Transduction
d) All of the above

Q6. Altered penicillin binding site is due to

a) Mec 2 gene
b) Mec 1 gene
c) Tet 1 gene
d) Inh A

Q7. Altered metabolic pathway confers resistance in

a) Quinolones
b) Sulpha drugs
c) Rifampicin
d) All of the above

Q8. Constitutive beta lactamase is present in

a) Gram positive bacteria
b) Gram negative bacteria
c) Both Aand B
d) None of these

Q9. Resistance against Aminoglycosides develops by
   a) Change in target site i.e. 30 S RNA
   b) Change in permeability and penetration
   c) Both A and B
   d) Efflux pump development

Q10. Resistance against Fluoroquinolones develops by
   a) Change in target site
   b) Change in permeability and penetration
   c) Both A and B
   d) Efflux pump development

Q11. Resistance against Fluoroquinolones develops by
   a) Mutation in Par C gene
   b) Mutation in Par E gene
   c) Mutation in Par C and Par E gene
   d) Mutation in Tet A gene

Q12. Example of acid resistant penicillin is
   a) Penicillin G
   b) Penicillin V
   c) Carbenicillin
   d) None of these

Q13. Example of penicillinase resistant penicillin is
   a) Cloxacillin
   b) Oxacillin
   c) Methicillin
   d) All of the above

Q14. Methicillin resistance occurs due to
   a) Altered enzyme activity
   b) Altered penicillin binding protein
   c) Both A and B
   d) None of these

Q15. MRSA is used for
   a) Monobactam resistant Staphylococcus aureus
   b) Minocycline resistant Staphylococcus aureus
   c) Methicillin resistant Staphylococcus aureus
   d) Midazolam resistant Staphylococcus aureus

Q16. Resistance against isoniazid is due to
   a) Mutation in Kat G
   b) Mutation in inh A
   c) Both A and B
   d) None of these

Q17. Tolerance and drug resistance can be a consequence of:

a) Drug dependence
b) Increased metabolic degradation
c) Depressed renal drug excretion
d) Activation of a drug after hepatic first-pass

Q18. Tolerance and drug resistance can be a consequence of:

a) Decreased metabolic degradation
b) Decreased renal tubular secretion
c) Change in receptors, loss of them or exhaustion of mediators
d) Increased receptor sensitivity

Q19. Tolerance develops because of:

a) Diminished absorption
b) Rapid excretion of a drug
c) Both of the above
d) None of the above

Q20. Thiazolidinediones act by:

a) Reducing the absorption of carbohydrate from the gut
b) Diminishing insulin resistance by increasing glucose uptake and metabolism in muscle and adipose tissues
c) Stimulating the beta islet cells of pancreas to produce insulin
d) All of the above

Q21. Alpha-glucosidase inhibitors act by:

a) Reducing the absorption of carbohydrate from the gut
b) Stimulating the beta islet cells of pancreas to produce insulin
c) Competitive inhibiting of intestinal α-ghucosidases & modulating the postprandial digestion and absorption of starch and disaccharides
d) Diminishing insulin resistance by increasing glucose uptake and metabolism in muscle and adipose tissues

Q22. Indication for 1,25-dihydroxyvitamin D3 (calcitriol) administration is:

a) Hypercalcemia of malignancy
b) Hypophosphatemia
c) Elevated skeletal turnover
d) Vitamin D resistance

Q23. Rational anti-microbial combination is used to:

a) Provide synergism when microorganisms are not effectively eradicated with a single agent alone
b) Provide broad coverage
c) Prevent the emergence of resistance
d) All of the above

Q24. Mechanisms of bacterial resistance to anti-microbial agents are the following, EXCEPT:

a) Active transport out of a microorganism or/and hydrolysis of an agent via enzymes produced by a microorganism

b) Modification of a drug's target

c) Enlarged uptake of the drug by a microorganism

d) Reduced uptake by a microorganism

Q25. Combined chemotherapy of tuberculosis is used to:

a) Increase mycobacterium drug-resistance

b) Decrease mycobacterium drug-resistance

c) Decrease the antimicrobal activity

d) Decrease the onset of antimycobacterial drugs biotransformation

Q26. Naturally occurring resistance present in gram negative bacteria against

a) Vancomycin b) Penicllin G

c) Both A and B d) None of these

Q27. Sudden change in genetic makeup is known as

a) Conjugation b) Transduction

c) Mutation d) Transformation

Q28. Neisseria acquired resistance against penicillin due to altered penicillin binding protein is developed by

a) Conjugation b) Transduction

c) Mutation d) Transformation

Q29. ……. involves a bacteriophage containing resistant genetic material for transmission of the resistant genetic material to the susceptible microorganism

a) Conjugation b) Transduction

c) Mutation d) Transformation

Q30. In the microbe of Enteriobacteriaceae and Pseudomonas sp following mode of developing resistance is common

a) Conjugation b) Transduction

c) Mutation d) Transformation

Q31. Altered binding site of 50 S RNA occurs in

a) Tetracycline b) Aminoglycosides

c) Erythromycin d) Penicillin

Q32. Phenomenon when the bacteria resistant to one antimicrobial resistance showing resistance to the other antimicrobial agent too, but the exposure to the latter has not occurred before.

a) Tolerance
b) Cross resistance
c) Tachyphylaxis
d) None of the above

Q33. Resistance to one antibiotic showing resistance to other and vice versa is also stands true in case of

a) Two way
b) Complete cross resistance
c) Both A and B
d) None of these

Q34. Example of two way cross resistance is

a) Gentamicin produces resistance to kanamycin
b) Neomycin and kanamycin
c) Both A and B
d) None of these

Q35. Resistance against one antibiotic confers resistance to other but resistance to latter could not produce resistance to the previous one is called as

a) Two way
b) Complete cross resistance
c) Partial resistance
d) None of these

Q36. Partial resistance is synonymous to

a) Incomplete cross resistance
b) One way resistance
c) Both A and B
d) None of these

Q37. Resistance to Gentamicin produces resistance to kanamycin and streptomycin but vice versa does not hold true is an example of

a) One way resistance
b) Two way
c) Complete cross resistance
d) None of these

Q38. In an anaesthetized dog, repeated intravenous injection of ephedrine shows the phenomenon of:

a) Anaphylaxis
b) Idiosyncrasy
c) Tachyphylaxis
d) Drug resistance

Q39. The monocomponent insulin preparations differ from the conventional preparations in the following respects except:

a) They cause less hypoglycaemic reactions
b) They are less allergenic
c) They cause less lipodystrophy
d) They are less liable to induce insulin resistance

Q40. Insulin resistance can be minimised by the use of:

a) Corticosteroids
b) Tolbutamide
c) Monocomponent/human insulin
d) Protamine

Q41. Which of the following is not a specific indication for the more expensive monocomponent/human insulins:

a) Sulfonylurea maintained diabetic posted for surgery
b) Type 1 diabetes mellitus
c) Insulin resistance
d) Pregnant diabetic

Q42. Select the drug which tends to reverse insulin resistance by increasing cellular glucose transporters:

a) Acarbose
b) Glibenclamide
c) Rosiglitazone
d) Prednisolone

Q43. The following is true of anti-H.pylori therapy except:

a) Resistance to any single antimicrobial drug develops rapidly
b) It is indicated in all patients of peptic ulcer
c) Concurrent suppression of gastric acid enhances efficacy of the regimen
d) Colloidal bismuth directly inhibits H.pylori but has poor patient acceptability

Q44. The following organism is notorious for developing antimicrobial resistance rapidly:

a) Treponema pallidum
b) Escherichia coli
c) Meningococcus
d) Streptococcus pyogenes

Q45. The most important mechanism of concurrent acquisition of multidrug resistance among bacteria is:

a) Mutation
b) Transduction
c) Conjugation
d) Transformation

Q46. Drug destroying type of bacterial resistance is important for the following antibiotics except:

a) Cephalosporins
b) Chloramphenicol
c) Aminoglycosides
d) Tetracyclines

Q47. Acquisition of inducible energy dependent efflux proteins by bacteria serves to:

a) Lyse host tissue
b) Confer antibiotic resistance
c) Secrete exotoxins
d) Enhance virulence

Q48. What is break point concentration of an antibiotic:

a) Concentration of the antibiotic which demarkates between sensitive and resistant bacteria

b) Concentration at which the antibiotic lyses the bacteria

c) Concentration of the antibiotic which overcomes bacterial resistance

d) Concentration at which a bacteriostatic antibiotic becomes bactericidal

Q49. Which of the following is not likely to be the cause of failure of antimicrobial therapy of an acute infection:

a) Failure to drain the pus

b) Improper selection of drug and dose

c) Acquisition of resistance during treatment

d) Uncontrolled diabetes mellitus

Q50. The fluoroquinolones have improved over nalidixic acid in the following respect(s):

a) Development of bacterial resistance against them is slow and infrequent

b) They have extended antimicrobial spectrum

c) They have higher antimicrobial potency

d) All of the above

Q51. The most common mechanism of development of resistance to fluoroquinolones is:

a) Acquisition of drug destroying enzyme

b) Acquisition of alternative metabolic pathway

c) Plasmid transfer

d) Chromosomal mutation altering affinity of target site

Q52. Important microbiological features of ciprofloxacin include the following except:

a) Marked suppression of intestinal anaerobes

b) MBC values close to MIC values

c) Slow development of resistance

d) Long postantibiotic effect

Q53. Indicate the disease in which penicillin G continues to be used as first line treatment in all cases (unless contraindicated), because the causative organism has not developed resistance so far:

a) Staphylococcal abscess

b) Gonorrhoea

c) Syphilis

d) Haemophillus influenzae meningitis

Q54. Bacteria develop tetracycline resistance by the following mechanisms except:

a) Synthesizing a 'protection protein' which interferes with binding of tetracycline to the target site

b) Losing tetracycline concentrating mechanisms

c) Actively pumping out tetracycline that has entered the cell

d) Elaborating tetracycline inactivating enzyme

Q55. The most important mechanism by which gram negative bacilli acquire chloramphenicol resistance is:

a) Lowered affinity of the bacterial ribosome for chloramphenicol

b) Decreased permeability into the bacterial cell

c) Acquisition of a plasmid encoded for chloramphenicol acetyl transferase

d) Switching over from ribosomal to mitochondrial protein synthesis

Q56. What is the most important reason for the restricted use of systemic chloramphenicol:

a) Emergence of chloramphenicol resistance

b) Its potential to cause bone marrow depression

c) Its potential to inhibit the metabolism of many drugs

d) Its potential to cause superinfections

Q57. The primary reason why chloramphenicol is not being used as the first line drug for typhoid fever in most areas is:

a) Delayed defervescence with chloramphenicol

b) Toxic potential of chloramphenicol

c) Delayed bacteriological cure with chloramphenicol

d) Spread of chloramphenicol resistance among S. typhi

Q58. Cross resistance among different members of the following class of antimicrobials is absent / incomplete or unidirectional:

a) Aminoglycosides b) Tetracyclines

c) Macrolides d) Both 'B' and 'C' are correct

Q59. The most important mechanism of bacterial resistance to an aminoglycoside antibiotic is:

a) Mutation reducing affinity of ribosomal protein for the antibiotic

b) Plasmid mediated acquisition of aminoglycoside conjugating enzyme

c) Mutational acquisition of aminoglycoside hydrolysing enzyme

d) Mutational loss of porin channels

Q60. The aminoglycoside antibiotic which is distinguished by its resistance to bacterial aminoglycoside inactivating enzymes is:

a) Sisomicin
b) Kanamycin
c) Tobramycin
d) Amikacin

Q61. The distinctive features of azithromycin include the following except: A.B. C.D.

a) Marked tissue distribution and intracellular penetration
b) Efficacy against organisms which have developed resistance to erythromycin
c) Long terminal elimination half-life
d) Low propensity to drug interactions due to inhibition of cytochrome P450 enzymes

Q62. What is true of isonicotinic acid hydrazide (INH):

a) Sensitive mycobacteria generate an active metabolite of INH through a catalaseperoxidase enzyme
b) An active transport mechanism concentrates INH inside sensitive mycobacteria
c) The most common mechanism of INH resistance is mutation in the target gene which encodes for a specific fatty acid synthase enzyme
d) Both 'A' and 'B' are correct

Q63. The primary reason for not using ethionamide as a first line antitubercular drug is:

a) It is only tuberculostatic and not tuberculocidal
b) It produces gastrointestinal intolerance and hepatitis
c) Ethionamide resistance has become widespread
d) It has to be given by injection

Q64. The most important reason for using a combination of chemotherapeutic agents in the treatment of tuberculosis is:

a) To obtain bactericidal effect
b) To broaden the spectrum of activity
c) To prevent development of resistance to the drugs
d) To reduce adverse effects of the drugs

Q65. Multidrug resistant (MDR) tuberculosis is defined as resistance to:

a) Any two or more antitubercular drugs
b) Isoniazid + any other antitubercular drug
c) Isoniazid + Rifampin + any one or more antitubercular drugs
d) All five first line antitubercular drugs

Q66. The multidrug therapy of leprosy is superior to monotherapy on the following account(s):

a) It prevents emergence of dapsone resistance

b) It shortens the total duration of drug therapy and improves compliance

c) It is effective in cases with primary dapsone resistance

d) All of the above

Q67. The following is true of multidrug therapy of leprosy except:

a) It has been highly successful in paucibacillary but not in multibacillary cases

b) No resistance to rifampin develops despite its use once a month

c) Relapse rate is very low in both paucibacillary and multibacillary cases

d) Prevalence of lepra reaction is not higher compared to dapsone monotherapy

Q68. Clonazepam produce its action by increasing the conductance of .... ions

a) Sodium

b) Potassium

c) Calcium

d) Chloride

Q69. Streptomycin interferes in protein synthesis by inhibiting the start codon

a) AUG

b) AGU

c) AAU

d) UUA

Q70. Which of the following is one of the most potent inhalant anaestetics

a) Methoxyflurane

b) Halothane

c) Isoflurane

d) Enflurane

Q71. Gabapentin acts

a) As $_{\text{GABAA}}$ agonist

b) As Precursor of GABA

c) By enhancing GABA release

d) By GABA independent mechanism

Q72. Which of the following drugs is antiprogestin

a) Gemeprost

b) Megestrol

c) Tamoxifen

d) Mifepristone

Q73. Zidovudine is effective against

a) HIV-1

b) HIV-1 and HIV-2

c) HIV-1, HIV-2 and HTLV-1

d) HIV-1, HIV-2, HTLV-I and HTLV-II

Q74. Identify the correct matching:

a) Yohimbine : alpha 2 adrenergic agonist

b) Xylazine : alpha 2 adrenergic antagonist

c) Glycine : Inhibitory neurotransmitter

d) 4-AP : Selective K-channel blocker

Q75. Thiamhenicol inhibits protein synthesis of the bacteria by binding with .... ribosomal sub unit

a) 30S

b) 50 S

c) 70 S

d) None of the above

Q76. Penicillins disrupt the formation of following peptide bond during its antibacterial action on bacteria:

a) D-alanine-D-aspartate

b) D-alanine- D-glutamate

c) D-Glycine-D-alanine

d) D-alanine-D-alanine

Q77. Erythromycin has antimicrobial activity as that of

a) Aminoglycosides

b) Natural penicillins

c) Tetracyclines

d) Quinolones

Q78. Local anesthetics produce:

a) Analgesia, amnesia, loss of consciousness

b) Blocking pain sensation without loss of consciousness

c) Alleviation of anxiety and pain with an altered level of consciousness

d) A stupor or somnolent state

Q79. A good local anesthetic agent shouldn't cause:

a) Local irritation and tissue damage

b) Fast onset and long duration of action

c) Systemic toxicity

d) Vasodilatation

Q80. Most local anesthetic agents consist of:

a) Amino group

b) Lipophylic group (frequently an aromatic ring)

c) Intermediate chain (commonly including an ester or amide)

d) All of the above

Q81. Which one of the following groups is responsible for the duration of the local anesthetic action?

a) Lipophylic group

b) Intermediate chain

c) Ionizable group

d) All of the above

Q82. Indicate the drug, which has greater potency of the local anesthetic action:

a) Procaine b) Mepivacaine

c) Bupivacaine d) Lidocaine

Q83. Which one of the following local anesthetics is an ester of benzoic acid?

a) Lidocaine b) Procaine

c) Cocaine d) Ropivacaine

Q84. Indicate the local anesthetic, which is an ester of paraaminobenzoic acid:

a) Mepivacaine b) Cocaine

c) Procaine d) Lidocaine

Q85. Indicate a reversible cholinesterase inhibitor:

a) Isoflurophate b) Carbochol

c) Physostigmine d) Parathion

Q86. Indicate a cholinesterase inhibitor, which has an additional direct nicotinic agonist effect:

a) Lobeline b) Neostigmine

c) Carbochol d) Edrophonium

Q87. Indicate the reversible cholinesterase inhibitor, which penetrates the blood-brain barrier:

a) Edrophonium b) Neostigmine

c) Physostigmine d) Piridostigmine

Q88. Which of the following drugs is both a muscarinic and nicotinic blocker?

a) Atropine b) Hexamethonium

c) Benztropine d) Succinylcholine

Q89. Atropine causes:

a) Mydriasis, a rise in intraocular pressure and cycloplegia

b) Miosis, a reduction in intraocular pressure and cyclospasm

c) Miosis, a rise in intraocular pressure and cycloplegia

d) Mydriasis, a rise in intraocular pressure and cyclospasm

Q90. Which competitive neuromuscular blocking agent could be used in patients with renal failure?

a) Atracurium b) Pipecuronium

c) Succinylcholine d) Doxacurium

Q91. Indicate the indirect-acting adrenoreceptor blocking drug

a) Prazosin
b) Tolazoline
c) Carvedilol
d) Reserpine

Q92. Compared with phentolamine, prazosin has all of the following features except.

a) Irreversible blockade of alfa receptors
b) Highly selective for alfa1 receptors
c) The relative absence of tachycardia
d) Persistent block of alfa1 receptors

Q93. Subtype-selective alfa1 receptor antagonists such as tamsulosin, terazosin, alfusosin are efficacious in:

a) Benign prostatic hyperplasia (BPH)
b) Hyperthyroidism
c) Cardiac arrhythmias
d) Asthma

Q94. Imidazopyridines are:

a) Partial agonists at brain 5-TH1A receptors
b) Competitive antagonists of BZ receptors
c) Selective agonists of the BZ1 (omega1) subtype of BZ receptors
d) Nonselective agonists of both BZ1 and BZ2 receptor subtypes

Q95. Which of the following hypnotic agents is able to interact with both BZ1 and BZ2 receptor subtypes?

a) Zaleplon
b) Phenobarbital
c) Flurazepam
d) Zolpidem

Q96. Indicate the agent, which interferes with GABA binding

a) Zolpidem
b) Thiopental
c) Bicuculline
d) Flurazepam

Q97. Which of the following agents blocks the chloride channel directly?

a) Zaleplon
b) Picrotoxin
c) Flumazenil
d) Secobarbital

Q98. Mechanism of action of cyclosporine A is:

a) Complement-mediated cytolysis of T lymphocytes
b) ADCC towards T lymphocytes
c) Compete for Fc receptors with autoantibodies
d) Inhibits calcineurin

Q99. Mechanism of action of I.V. IgG preparation is:

a) Complement-mediated cytolysis of T lymphocytes
b) Compete for Fc receptors with autoantibodies
c) Inhibits CD3 receptor
d) Inhibits calcineurin

Q100. Mechanism of action of sirolimus (rapamycin) is:

a) Modulation of CD3 receptor from the cell surface
b) ADCC towards T lymphocytes
c) Inhibits calcineurin
d) Anti-idiotype antibodies against autoantibodies

Q101. Mechanism of action of levamisole is:

a) Substitution for patient's defiecient immunoglobulins
b) Increase the number of T-cells
c) Complement-mediated cytolysis of T lymphocytes
d) Inhibits CD3 receptor

Q102. Mechanism of action of calcitonin is:

a) Inhibits macrophages
b) Inhibits hydroxyapatite crystal formation, aggregation, and dissolution
c) Raises intracellular cAMP in osteoclasts
d) Activates bone resorption

Q103. Mechanisms of bacterial resistance to anti-microbial agents are the following, except.

a) Active transport out of a microorganism or/and hydrolysis of an agent via enzymes produced by a microorganism
b) Enlarged uptake of the drug by a microorganism
c) Modification of a drug's target
d) Reduced uptake by a microorganism

Q104. Mechanism of penicillins' antibacterial effect is:

a) Inhibition of beta-lactamase in the bacterial cell
b) Inhibition of transpeptidation in the bacterial cell wall
c) Activation of endogenous proteases, that destroy bacterial cell wall

Q105. Activation of endogenous phospholipases, which leads to alteration of cell membrane permeability

a) Mechanism of Amphotericin B action is: Inhibition of cell wall synthesis

b) Inhibition of DNA synthesis
c) Inhibition of fungal protein synthesis
d) Alteration of cell membrane permeability

Q106. Mechanism of Trimethoprim' action is:
a) Inhibition of DNA gyrase
b) Inhibition of dihydropteroate synthase
c) Inhibition of dihydropteroate reductase
d) Inhibition of cyclooxygenase

Q107. Mechanism of Izoniazid action is:
a) Inhibition of protein synthesis
b) Inhibition of mycolic acids synthesis
c) Inhibition of RNA synthesis
d) Inhibition of ADP synthesis

Q108. Mechanism of Rifampin action is:
a) Inhibition of mycolic acids synthesis
b) Inhibition of cAMP synthesis
c) Inhibition of DNA dependent RNA polymerase
d) Inhibition of topoisomerase II

Q109. Mechanism of Cycloserine action is:
a) Inhibition of mycolic acids synthesis
b) Inhibition of cell wall synthesis
c) Inhibition of RNA synthesis
d) Inhibition of pyridoxalphosphate synthesis

Q110. Mechanism of aminosalicylic acid action is:
a) Inhibition of DNA gyrase
b) Inhibition of DNA dependent RNA polymerase
c) Inhibition of folate synthesis
d) Inhibition of mycolic acids synthesis

Q,111. Action mechanism of methotrexate is:
a) Catabolic depletion of serum asparagine
b) Inhibition of dihydrofolate reductase
c) Activation of cell differentiation
d) All of the above

Answer Keys

1	a	2	d	3	d	4	b	5	a	6	b	7	b
8	b	9	c	10	d	11	c	12	b	13	d	14	b
15	c	16	c	17	b	18	c	19	d	20	b	21	c
22	b	23	d	24	c	25	b	26	c	27	c	28	d
29	b	30	a	31	c	32	b	33	c	34	b	35	c
36	c	37	a	38	c	39	a	40	c	41	b	42	c
43	b	44	b	45	c	46	d	47	b	48	a	49	c
50	d	51	d	52	a	53	c	54	??	55	c	56	b
57	d	58	a	59	b	60	d	61	b	62	d	63	b
64	c	65	c	66	d	67	a	68	d	69	a	70	a
71	c	72	d	73	d	74	c	75	c	76	d	77	b
78	b	79	b	80	d	81	b	82	c	83	c	84	c
85	c	86	b	87	c	88	c	89	a	90	a	91	d
92	a	93	a	94	c	95	c	96	c	97	b	98	d
99	b	100	c	101	b	102	c	103	b	104	b	105	d
106	c	107	b	108	c	109	b	110	c	111	b		

3

Chemotherapy

Amit Shukla, Soumen Choudhury and Atul Prakash

Q1. The nitrogen atom of amino group in basic structure of sulfonamide is designated as

a) N1 b) N2

c) N3 d) N4

Q2. Trimethoprim produces synergistic action by interfering the synthesis of

a) Dihydrofolic acid from tetrahydrofolic acid

b) Tetrahydrofolic acid from dihydrofolic acid

c) mRNA from DNA

d) tRNA from DNA

Q3. An ophthalmic sulfonamide is

a) Sulfaguinidine b) Sulfadimidine

c) Sulfadimethoxine d) Sodium sulfacetamide

Q4. Toxicity of sulfonamide is well characterized by

a) Nephrotoxicity b) Hepatotoxicity

c) Reproductive toxicity d) All of the above.

Q5. Penicillins disrupt the formation of following peptide bond during its antibacterial action on bacteria:

a) D-alanine-D-aspartate b) D-alanine- D-glutamate

c) D-Glycine-D-alanine d) D-alanine-D-alanine

Q6. A β-lactamase resistant penicillin is

a) Penicillin-G b) Penicillin-V

c) Oxacillin d) Benzyl penicillin

Q7. Oxytetracycline is effective against

a) Gram + organisms b) Gram- organisms

c) Anaplasm d) All of the above

Q8. Toxicity of aminoglycosides is characterized by

a) Ototoxicity b) Renal toxicity

c) Neuromuscular paralysis d) All of the above.

Q9. Aminoglycoside resistant to aminoglycoside inactivating enzyme is

a) Streptomycin b) Gentamicin

c) Kanamycin d) Amikacin

Q10. Thiamhenicol inhibits protein synthesis of the bacteria by binding with .... ribosomal sub unit

a) 30S b) 50 S

c) 70 S d) None of the above

Q11. A fluorine containing derivative of chloramphenicol is

a) Thiamphenicol b) Florfenicol

c) Chloramphenicol d) None of the above.

Q12. Erythromycin has antimicrobial activity as that of

a) Natural penicillins b) Aminoglycosides

c) Tetracyclines d) Quinolones.

Q13. Identify the incorrect pair

a) Colistin-Polypeptide b) Erythromycin-Macrolide

c) Cephalosporin-β-lactam d) Quinolones-Cyclohexanes

Q14. *In E.coli*, topoisomerase IV, the target enzyme of the quinolones is encoded by the genes

a) parC and parE b) gyrA and gyrB

c) par C and gyr A d) parE and gyrB

Q15. Antibacterial activity of nitrofurans is attributed to

a) C-2 substitution b) C-3 substitution

c) 5-nitro group d) Replacement of 0-atom by S-atom

Q16. Miconazole inhibits the synthesis of fungal

a) Ergesterol b) Cholesterol

c) DNA replication d) Proteins

Q17. Identify the incorrect pair

a) Florey -Penicillins

b) Woods and Fildes-Sulphonamide

c) Shatz and Bugie-Streptomycin

d) Duggar-Oxytetracycline

Q18. Piperazine produces its action by

a) Neuromuscular paralysis b) Inhibiting ATP synthesis

c) Inhibition of glycolysis d) None of the above

Q19. Prazequantel is used as

a) Antitrematodal drug
b) Anticestodal drug
c) Antinematodal drug
d) Antimalarial drug

Q20. Which of the following is polyether antibiotic obtained as fermentation product of Streptomyces :

a) Lasalocid
b) Maduremicin
c) Narasin
d) All of the above

Q21. The superinfection is highest with the therapy of .... antimicrobials.

a) Broad spectrum
b) Narrow spectrum
c) combined narrow and broad spectrum
d) Combined broad spectrum and third generation cephalosporins

Q22. Ivermectin is agonist of

a) PABA
b) GABA
c) Glycine
d) Aspartate

Q23. Penicillins disrupt bacterial cell wall synthesis by inhibiting

a) DNAgyrase
b) DNA polymerase
c) PBPs enzymes
d) None of the above

Q24. Which of the following antibiotics have similar binding site in the susceptible bacteria to produce their antibacterial action:

a) Penicillins and tetracyclines
b) Tetracyclines and aminoglycosides
c) Aminoglycosides and quinolones
d) Quinolones and macrolides

Q25. Aminoglycosides are very poorly absorbed following .... route of administration

a) Oral
b) Intramuscular
c) Intravenous
d) Intraperitoneal

Q26. Streptomycin interferes in protein synthesis by inhibiting the start codon

a) AUG
b) AAU
c) UUA
d) AGU

Q27. The binding site of aminoglycosides in microbial ribosome consists of

a) 16 proteins and 16S molecule of RNA
b) 21 proteins and a 16S molecule of RNA

c) 21 proteins and 21S molecule of RNA

d) 16 proteins and 21S molecule of RNA

Q28. Methenamine acts as urinary antiseptic due to formation of:

a) Ethanol b) Methanol

c) Formaldehyde d) Acetaldehyde

Q29. Levamisol removes nematodes by producing .... action:

a) Cholinomimetic b) Antiglycolytic

c) Inhibition of electron transport d) Neuromuscular blockade

Q30. Monesin is a .... ionophore antibiotic.

a) Monovalent Polyether b) Divalent Polyether

c) Polyvalent Polyether d) Glycopeptide

Q31. A penicillin effective against Pseudomonas is

a) Carboxypenicillins b) Ureidopenicillins

c) Both of the above d) None of the above

Q32. Which of the following is bacteriostatic in therapeutic doses:

a) Tetracyclines b) Fluoroquinolones

c) Penicillins d) Aminoglycosides

Q33. Quinolones inhibit the enzyme (s) to produce antibacterial action:

a) Topoisomerase IV

b) DNA gyrase

c) DNA dependent RNA polymerase

d) Topoisomerase IV and DNA gyrase

Q34. Which of the following aminoglycosides has extensive plasma albumin binding:

a) Kanamycin b) Gentamicin

c) Tobramycin d) Streptomycin

Q35. Source of chlortetracycline is

a) Strepto. venezuele b) Strepto. tenebrarius

c) Strepto. aureofaciens d) Strepto. orientalis

Q36. Tetracyclines block the tRNA binding to .... on ribosome to produce antibacterial action:

a) A site b) P site

c) Both A and B sites d) Polypeptide chain

Q37. .... a potent anticancer cytokinin.

a) Interferon
b) Penicillin binding proteins
c) Pork nucleotides
d) All of the above

Q38. A benzene sulfonamide antitrematodal agent

a) Sulfamylon
b) Rafoxanide
c) Clioxanide
d) Clorsulon

Q39. Tetracycline effective in STD is

a) Minocycline
b) Demeclocyline
c) Doxycycline
d) Oxytetracycline

Q40. A bitch treated indiscriminately with streptomycin during pregnancy is most likely to have progeny with irreversible defects of

a) Vision
b) Hearing
c) Olfactory sensation
d) Taste sensation

Q41. Which of the following depicts the correct order of absorption of tetracyclines from GIT in canines:

a) Minocycline> Oxytetracycline >Dopxycycline>Chlortetracycline
b) Minocycline>Dopxycycline>Chlortetracycline> Oxytetracycline
c) Dopxycycline>Minocycline > Oxytetracycline>Chlortetracycline
d) Minocycline>Dopxycycline> Oxytetracycline>Chlortetracycline

Q42. Drug of choice to treat penicillin induced anaphylactic shock is

a) Dexamethasone
b) Chlorpheniramine
c) Epinephrine
d) Dobutamine

Q43. .... is an antitrypanosomal drug

a) Suramin
b) Homidinium
c) Buparvaquone
d) Parvaquone

Q44. Which of the following is/are drug(s) for amoebiasis in dogs:

a) Metronidazole
b) Diloxonide furoate
c) Iodoquinol
d) All of the above

Q45. Rafoxanide is ideally combined with .... for treating fascioliasis in cattle.

a) Nitroxinil
b) Oxyclozanide
c) Diamfenetide
d) Niclofolan

Q46. Sulphonamides produce weak shelled eggs in laying birds by inhibiting

a) Dihydrofolate synthetase
b) Carbonic anhydrase
c) Dihydrofolate reductase
d) Superoxide dismutase

Q47. Which of the following anthelmintic inhibits glycolysis:

a) Thiacetarsamide
b) Poassium antimony tartarate
c) Stibophen
d) All of the above

Q48. Zidovudine is effective against

a) HIV-1
b) HIV-1 and HIV-2
c) HIV-1, HIV-2 and HTLV-1
d) HIV-1, HIV-2, HTLV-I and HTLV-II

Q49. Oxyclozanide producesantitrematodal action by inhibiting synthesis of

a) ATP
b) Glycogen
c) Ergesterol
d) NADP

Q50. Amoxicillin has no synergism with:

a) Clavulanic acid
b) Diamonopyrimidines
c) Tazobactum
d) Sulbactum

Answer Keys

1	a	2	b	3	d	4	a	5	d	6	c	7	d
8	d	9	d	10	c	11	b	12	a	13	d	14	a
15	c	16	a	17	d	18	a	19	b	20	d	21	d
22	c	23	c	24	b	25	a	26	a	27	b	28	c
29	a	30	a	31	c	32	a	33	d	34	d	35	c
36	a	37	a	38	d	39	c	40	b	41	d	42	c
43	a	44	d	45	c	46	b	47	d	48	d	49	a
50	b												

4

Toxicopathology

Sakshi Tiwari and Amit Shukla

Q1. Glucuronide conjugation occurs in the compounds with following functional groups:

a) Hydroxyl b) Carboxyl
c) Amino d) All above

Q2. Which of the following is not a chelating agent:

a) Dimercaprol b) Penicillamine
c) Dimethyl sulfoxide d) Ca-EDTA

Q3. Most appropriate treatment of acute organophosphate toxicity:

a) Atropine alone b) 2-PAM alone
c) Atropine + 2-PAM d) Sod. Thiosulphate alone

Q4. Acute urea poisoning in cattle occurs due to excessive ruminal formation of:

a) Propionic acid b) Ammonia
c) Methane d) Butyric acid

Q5. Cyanides inhibit following enzyme to produce their toxicity:

a) Glucuronyl transferase b) Cytochrome oxidase
c) Acetylcholine esterase d) Sulfurtransferase

Q6. Lantana camara causes .... in animals.

a) Primary photosensitization b) Secondary photosensitization
c) Ophthalmic disorders d) Red urine

Q7. Ochratoxins are lethal due to .... in animals.

a) Hepatotoxicity b) Nephrotoxicity
c) Neurotoxicity d) Cardiac toxicity

Q8. Synthetic pyrethroids increase primarily the conductance of .... channels.

a) Calcium b) Potessium
c) Sodium d) Chloride

Q9. Which of the pair is incorrect:

a) Arsenic – Alpha Lipoic acid.

b) Organophosphatic – Acetyl cholinesterase

c) Hydrocyanamic acid – Cyt P_{450}

d) Sodium fluoroacetate – Succinate dehydrogenase

Q10. EDRF has been identifies as

a) Nitric oxide b) Sulfan sulfur

c) cAMP d) cGMP

Q11. BAL contains .... thiol groups.

a) One b) Two

c) Three d) Four

Q12. Identify the incorrect pair:

a) Dimercaprol- Mercury b) Deferoxamine - Copper

c) CaNa2EDTA - Lead d) Dithizone - Thallium

Q13. Methylene blue is antidote of .... poisoning in animals:

a) Cyanides b) Warfarins

c) DNOC d) Lindane

Q14. Rhodanese is also known as:

a) Thiosulfate sulfur transferase b) Glutathions-s-transferase

c) Sulfan sulfur d) Super oxide dismutase

Q15. Cyanide radicals bind to .... of the cytochrome complex:

a) Only ferric iron b) Only cupric copper

c) Both of the above d) None of the above

Q16. For specimen collection for diagnosis of poisoning, which pair is incorrect

a) Aflatoxicosis - Kidney b) Antimory - Liver

c) Fluorosis - Bones d) Lead - Blood

Q17. Subacute toxicity studies should be carried out atleast for:

a) 60 days b) 90 days

c) 45 days d) 21 days

Q18. Poisons hindering oxygen transport :

a) Nitrates b) Chlorates

c) Copper d) All of the above

Q19. Poison hindering oxygen uptake from alveoli.

a) ANTU b) Sulfur oxide

c) Paraquat d) All of the above

Q20. Form(s) of lead present in petrol:

a) Tetraethyl lead
b) Tetraethyl and Tetramethyl lead
c) Lead acetate
d) Lead acetate and tetramethyl lead.

Q21. Which of the following is/are known as greenhouse gas(s):

a) Methane
b) Carbondioxide
c) Carbonmonoxide
d) All of the above.

Q22. Bone marrow is the target organ of toxicity of:

a) Chloramphenicol
b) Radiation
c) Chloramphenicol and radiation
d) None of the above

Q23. Which of the following pairs of toxicants-sinking sites is incorrect:

a) Fluorides and lead – Bones
b) Arsenic – Liver
c) Halogenated Hydrocarbons – Adipose tissue
d) Iodine – thyroid gland

Q24. Following enzyme is not involved in phase II metabolism of toxicants.

a) Glucuronyl transferase
b) Glutathione –S-transferase
c) Sulfotransferase
d) Cyt P450 dependent monoopxygenase

Q25. Which of the following is not defined as a xenobiotic:

a) Polypeptide vaccines
b) Industrial effluents
c) Agrochemicals
d) Vit B-comlex

Q26. Co-factor required for methylation of xenobiotics is:

a) S-adenosyl methionine
b) Glutathione
c) Acetyl Co A
d) 3’ Phospho adenosine 5’ phosphosulphate

Q27. Metabolism by N-acelytation occurs in xenobiotics containing--- groups.

a) R-NH2
b) R-NH2 and R-NH-NH2
c) R-OH group
d) R-SH group

Q28. UDPGA provide endogenous substrate for:

a) Glucuronidation
b) Sulfation
c) Acelytation
d) Methylation

Q29. Garlic odour of expired air indicates poisoning of ....

a) Phosphorus b) Lead
c) Cyanide d) Chlorate

Q30. Characteristic mice smell is indication of poisoning of :

a) Cyanide b) Hemlock
c) OP compound d) ANTU

Q31. CS- syndrome is observed in toxicity of

a) Type I synthetic pyrethroid b) Type II synthetic pyrethroid
c) Organochlorine d) Carbamates

Q32. Madness syndrome with hyperthermia is characteristic in poisoning of:

a) Synthetic pyrethroides b) Organophosphatess
c) Carbamates d) Organochlorine

Q33. Bright red blood is an indication in toxicity of :

a) Nitrate b) Chlorate
c) Dinitro phenols d) Cyanide

Q34. Sodium channels are the targets for action of:

a) Synthetic pyrethroid b) Organochlorine
c) Local anaesthetics d) All of the above

Q35. Dose of absorbed radiation is expressed in unit of :

a) Gay b) Joul
c) Erg d) Dalton

Q36. AChE is a biomarker for toxicity of :

a) Organophosphate
b) Carbamate and organophosphate
c) Synthetic pyrethroid
d) Carbamate and organochlorine

Q37. Niigata Disease occurred due to contamination of water with

a) DDE b) Methyl mercury
c) Calomel d) Mersalyl

Q38. Which of the following known as ozone depleting substance (ODS):

a) Freons only
b) Halons only
c) Freons and halones
d) Freons, halones, carbon tetrachloride and methyl chloroform

Q39. Antidote to methanol poisoning is:

a) Atropine
b) Sod. Thiosulphate
c) Ethanol
d) Dimercaprol

Q40. Identify incorrect matching:

a) Alkali disease –Elongated hooves
b) Pacing disease- Spectacled eye
c) Haemolytic crisis-Copper
d) KD effect-Carbamates

Q41. Identify incorrect pair of herbicides:

a) Substituted urea compounds-Dalapon
b) Phenoxyacetic acid -2,4-D
c) Heterocyclic compounds-Atrazine
d) Bipyridinium -Diquat

Q42. Methemoglobin is converted to hemoglobin in blood by:

a) Sulfer transferases
b) Dia phorase I
c) Diaphorases I and II
d) Glutathion-S-Transferase

Q43. Nitric oxide causes vasodialation by a stimulating the enzyme:

a) Adenylyl cyclase
b) Guanylyl cyclase
c) Phospholipase A
d) Phospholipase C

Q44. Identify pair showing incorrect matching of plant – glycoside:

a) Linseed meal – Linamarin
b) Lotus sp. – Lotusin
c) Sorghum - Dhurrin
d) Bitter almonds – Gossipol

Q45. Which of the following is/are appropriate for treating cyanide toxicity in cattle:

a) Sod.nitrite, sod thiosulphate and vinegar
b) Sod Thiosulphate only
c) Sod Nitrite and sod thiosulphate only
d) Sod Nitrite only

Q46. Oleander nerium cardiotoxins interfere with :

a) Na+-K+ ATPase
b) Protein kinase
c) Adenyl cyclase
d) Guanyl cyclase

Q47. Strychnine produces toxicity by inhibiting:

a) GABA receptors
b) Glycine receptors
c) Muscarinic receptors
d) Nicotinic receptor

Q48. Toxic principle present in caster oil cake:

a) Ricin I
b) Ricin I and II
c) Abrin
d) Gossipole

Q49. Which of the statements is inappropriate about AChE reactivator:

a) Not used in carbamate poisoning
b) Used only in initial stage of OP poisoning
c) 2-PAM has longer half life than DAM
d) Not used after ageing of AChE

Q50. Which is most toxic aflatoxin:

a) Aflatoxin B1
b) Aflatoxin G1
c) Aflatoxin B2
d) Aflatoxin G2

Answer Keys

1	d	2	c	3	c	4	b	5	b	6	b	7	b
8	c	9	d	10	a	11	b	12	b	13	c	14	a
15	c	16	a	17	c	18	d	19	d	20	b	21	d
22	c	23	b	24	d	25	d	26	a	27	b	28	a
29	a	30	b	31	b	32	d	33	d	34	d	35	a
36	b	37	b	38	d	39	c	40	d	41	a	42	c
43	b	44	d	45	a	46	a	47	b	48	b	49	c
50	a												

5

Practice Set-1

Amit Shukla, Soumen Choudhury and Atul Prakash

Q1. Penicillins disrupt the formation of following peptide bond during its antibacterial action on bacteria:

a) D-alanine-D-aspartate
b) D-alanine- D-glutamate
c) D-Glycine-D-alanine
d) D-alanine-D-alanine

Q2. In *E.coli*, topoisomerase IV, the target enzyme of the quinolones is encoded by the genes

a) parC and parE
b) gyrA and gyrB
c) par C and gyr A
d) parE and gyrB

Q3. Streptomycin interferes in protein synthesis by inhibiting the start codon

a) AUG
b) AAU
c) UUA
d) AGU

Q4. The binding site of aminoglycosides in microbial ribosome consists of

a) 16 proteins and 16S molecule of RNA
b) 21 proteins and a 16S molecule of RNA
c) 21 proteins and 21S molecule of RNA
d) 16 proteins and 21S molecule of RNA

Q5. Tetracyclines block the tRNA binding to .... on ribosome to produce antibacterial action:

a) A site
b) P site
c) Both A and B sites
d) Polypeptide chain

Q6. Which of the following depicts the correct order of absorption of tetracyclines from GIT in canines:

a) Minocycline> Oxytetracycline >Dopxycycline>Chlortetracycline
b) Minocycline>Dopxycycline>Chlortetracycline> Oxytetracycline
c) Dopxycycline>Minocycline > Oxytetracycline>Chlortetracycline
d) Minocycline>Dopxycycline> Oxytetracycline>Chlortetracycline

Q7. Drug of choice to treat penicillin induced anaphylactic shock is

a) Dexamethasone b) Chlorpheniramine

c) Epinephrine d) Dobutamine

Q8. Zidovudine is effective against

a) HIV-1

b) HIV-1 and HIV-2

c) HIV-1, HIV-2 and HTLV-1

d) HIV-1, HIV-2, HTLV-I and HTLV-II

Q9. EDRF has been identifies as

a) Nitric oxide b) Sulfan sulfur

c) cAMP d) cGMP

Q10. Rhodanese is also known as:

a) Thiosulfate sulfur transferase b) Glutathions-s-transferase

c) Sulfan sulfur d) Super oxide dismutase

Q11. Following enzyme is not involved in phase II metabolism of toxicants.

a) Glucuronyl transferase

b) Glutathione –S-transferase

c) Sulfotransferase

d) Cyt P450 dependent monoopxygenase

Q12. Co-factor required for methylation of xenobiotics is:

a) S-adenosyl methionine

b) Glutathione

c) Acetyl Co A

d) 3' Phospho adenosine 5' phosphosulphate

Q13. Metabolism by N-acelytation occurs in xenobiotics containing .... groups.

a) R-NH2 b) R-NH2 and R-NH-NH2

c) R-OH group d) R-SH group

Q14. Dose of absorbed radiation is expressed in unit of :

a) Gay b) Joul

c) Erg d) Dalton

Q15. Which of the following known as ozone depleting substance (ODS):

a) Freons only

b) Halons only

c) Freons and halones

d) Freons, halones, carbon tetrachloride and methyl chloroform

Q16. Nitric oxide causes vasodialation by a stimulating the enzyme:

a) Adenylyl cyclase
b) Guanylyl cyclase
c) Phospholipase A
d) Phospholipase C

Q17. Oleander nerium cardiotoxins interfere with :

a) Na+-K+ ATPase
b) Protein kinase
c) Adenyl cyclase
d) Guanyl cyclase

Q18. Which of the following is not required for oxidative reaction by MFOs:

a) Cyt P_{450}
b) Oxygen
c) Reduced NADP
d) H_2O

Q19. The drug with the following therapeutic index will have widest margin of safety:

a) 14.0
b) 15.0
c) 13.0
d) 14.5

Q20. Which of the following is one of the most potent inhalant anaestetics:

a) Methoxyflurane
b) Isoflurane
c) Enflurane
d) Halothane

Q21. Identify the correct matching:

a) Xylazine : α2 adrenergic antagonist
b) Yohimbine : α2 adrenergic agonist
c) Glycine : Inhibitory neurotransmitter
d) 4-AP : Selective K-channel blocker

Q22. Benzodiazepines produce their action by increasing the conductance of .... ions.

a) Sodium
b) Potassium
c) Chloride
d) All of the above

Q23. Disopyramide, an antiarrhythmic agent, act as a :

a) A calcium channel blocker
b) A cell membrane stabilizer
c) An adrenolytic agent.
d) An agent prolonging action potential duration

Q24. The site of action of thiazide diuretics is

a) Distal convoluted tubules
b) Proximal convoluted tubules

c) Ascending loop of Henle

d) descending loop of Henle

Q25. Which of the following is a systemic anticoagulant to produce immediate action:

a) Sod. fluoride b) Warfarin

c) Heparin d) Editate sodium

Q26. Low doses of aspirin prolong bleeding time by inhibiting the following in platelets:

a) Thromboxane A2 b) 5-HT

c) PAF d) PG E2

Q27. Which of the following drugs is antiprogestin:

a) Gemeprost b) Megestrol

c) Mifepristone d) Tamoxifen

Q28. Which of the following drugs selectively blocksAT1 receptors:

a) Ramipril b) Lovastatin

c) Candesartan d) Sumatriptan

Q29. Which of the following drug (s) produce(s) extrapyramidal side effects:

a) Metoclopramidel b) Cisapride

c) Domperidone d) All of the above

Q30. Gabapentin acts :

a) As GABAA agonist

b) As Precursor of GABA

c) By enhancing GABA release

d) By GABA independent mechanism

Answer Keys

1	d	2	a	3	a	4	b	5	a	6	d	7	c
8	d	9	a	10	a	11	d	12	a	13	b	14	a
15	d	16	b	17	a	18	d	19	b	20	a	21	c
22	c	23	b	24	a	25	c	26	a	27	c	28	c
29	a	30	c										

6

Practice Set-2

Soumen Choudhury, Amit Shukla and Atul Prakash

Q1. The exit of drug from C.S.F. irrespective of its entry is by:

a) Passive diffusion
b) Filtration
c) Facilitated diffusion
d) All of the above.

Q2. The number of half-lives required to reach 99% of steady state concentration is:

a) 3.3
b) 4.4
c) 5.5
d) 6.6

Q3. The receptor concept was first introduced by:

a) J.N. Langely in 1878
b) Simonis in 1964
c) Paul Ehrlich in 1926
d) Wakcsman in 1826

Q4. Characteristic untoward symptoms unrelated to dose or pharmacodynamic poroperty of the drug is called

a) Idiosyncrasy
b) Toxic effect
c) Hypersensitivity
d) Intolerance

Q5. If the combined effect of two drugs acting by the same mechanism is equal to the algebraic sum of their individual effect, it is called as:

a) Antagonism
b) Potentiation
c) Additive effect
d) None of the above

Q6. For increasing the excretion of weakly acidic drugs, urine should be made:

a) Alkaline
b) Acidic
c) At neutral pH
d) pH does not affect the urinary excretion of acidic drugs

Q7. Drugs that have steep does response curve and a small therapeutic ratio are following, except.

a) Antiarrhythmic drugs
b) Anticancer drugs
c) Antiepileptic drugs
d) Penicillins

Q8. First order kinetics of the drugs is called when:

a) A constant fraction of the drug is removed in per unit time

b) A constant amount of the drug is removed in per unit time

c) Total amount of the drug is removed in one hour

d) Total amount of the drug is removed in first passage through the kidneys

Q9. 'Aging' is a term applied to 'phosphorylated acetylcholinestrase' enzyme, which is due to:

a) Loss of one of the alkoxy gr.

b) Acquisition of negative charge

c) Both of the above

d) None of the above

Q10. Hydrolysis of cAMP is catalysed by several phosphodiesterases which may be activated by:

a) Calcium b) Calmodulin

c) Both of the above d) None of the above.

Q11. Acetylcholinesterase anionic subsite inhibitor is:

a) Tetramethyl ammonium ion b) Edrophonium

c) Both of the above d) None of the above

Q12. Hemicholinium modifies Ach metabolism by

a) Interfering with the destruction of Ach

b) Interfering with the release of Ach

c) Interfering with the synthesis of Ach

d) Promoting release of Ach.

Q13. Which of the following is the rate limiting enzyme in the synthesis of norepinephrine?

a) Tyrosine hydroxylase b) Dopa decarboxylase

c) Dopamine beta hydroxylase d) N-methyl transferase

Q14. In myasthenia gravis the treatment of cholinergic crisis should be :

a) Withdrawing all anticholinesterase medication

b) IV injection of atropine

c) IV injection of pralidoxime

d) All of the above.

Q15. Hyoscine (scopolamine) differs from atropine in that:

a) It has CNS depressant action

b) Mydriasis is longer than atropine

c) It has got some adrenergic property also
d) All of the above.

Q16. In which of the following organs both α and β adrenergic stimulation produce the same effect
a) Blood vessels
b) Intestine
c) Uterus
d) Bronchial muscle

Q17. Adrenaline enhances intracellular cyclic AMP by an action of
a) α receptors
b) β receptors
c) α and β receptors both
d) None of the two

Q18. Which of the following drugs is useful in the treatment of urinary retention?
a) Atropine
b) Epinephrine
c) Carbachol
d) Bethanecol

Q19. The site of action of COMT on catecholamine is
a) Ethanolamine
b) Catechol nucleus
c) Carboxy group
d) Hydroxy group

Q20. D2 (dopamine) agonist
a) Sarlasin
b) Bromocriptine
c) Renin
d) Aldosterone

Q21. Long acting alpha-1 blocker.
a) Doxazocin
b) Prazosin
c) Xylazine
d) None of them

Q22. Phosphodiesterase is found in.
a) Heart
b) Skeletal muscle
c) Fat
d) Lungs

Q23. It inhibits release of substance P
a) Theophylline
b) Naloxone
c) Morphnine
d) Camphor

Q24. Extra chromosomal resistance can take place by.
a) Transduction
b) Transformation
c) Conjugation
d) All of the above

Q25. If FIC (fractional inhibitory concentration) index is less than one then the combination of drug is:
a) Synergistic
b) Additive
c) Antagonistic
d) None of the above

Q26. Solubility of sulfonamides depends upon.

a) pH
b) Its salt
c) Metabolites
d) All of the above

Q27. Substitution in sulfonamide structures for systemic effects are generally made at:

a) N1 position
b) N4 position
c) Both of the above
d) None of the above

Q28. The basis of selective toxicity of trimethoprim is that

a) Mammalian cells have different metabolic pathway for the synthesis of tetrahydrofolate
b) Mammalian cells possess transport mechanism that enable them to absorb tetrahydrofolate in preformed state.
c) The enzyme required for the formation of tetrahydrofolate is less inhibitory in mammalian cell than the bacterial cell.
d) All of the above

Q29. Second generation quinolones are mainly effective against

a) Gram positive organism
b) Gram negative organism
c) Acid fast organism
d) All of the above.

Q30. Park nucleotide consists of

a) UDP acetyl muramyl pentapeptide
b) UDP acetyl glucosamine
c) Both of the above
d) None of the above

Q31. The volume of distribution of tetracycline is:

a) Equivalent to total body water
b) Less than the total body water
c) Greater than the total body water
d) None of the above.

Q32. Demeclocycline produces following kind of photosensitization.

a) Primary
b) Secondary
c) Both of the above
d) None of the above

Q33. Tetracycline inhibits protein synthesis in bacteria by

a) Distorting 30s ribosomal subunit so that correct message from m-RNA is not translated
b) Preventing binding of t-RNA to m-RNA.
c) Preventing binding of t-RNA on acceptor site in 50s ribosomal subunit.
d) All of the above

Q34. Volume of distribution of gentamicin is.

a) Less than the volume of extracellular fluid
b) Equal to volume of extracellur fluid

c) More than the volume of extracellular fluid
d) None of the above

Q35. Aminoglycosides inhibits protein synthesis in bacteria by
a) Preventing addition of amino acid to growing polypeptide chain
b) Causing misreading of codon on the m-RNA
c) Preventing binding of t-RNA to m-RNA
d) Preventing binding of t-RNA on acceptor site in 50s ribosomal subunit

Q36. The binding site of aminoglycoside on 30s ribosomal subunit consists of
a) 13 proteins and 16s ribosomal RNA
b) 3 proteins and 16s ribosoomal RNA
c) 1 protein and 16s ribosomal RNA
d) None of the above

Q37. Neuromuscular blockade by aminoglycosides can be overcome by
a) Administration of choline esterase inhibitors
b) Administration of i.v.calcium preparation
c) Both of the above
d) None of the above

Q38. Macrolides chemically consists of
a) Amino sugars linked by glycoside linkage
b) Lactone ring with two sugars
c) Both of the above.
d) None of the above

Q39. The mechanism of action of Griseofulvin is that it
a) Interferes with the cell membrane permeability
b) Interferes with fungal mitosis
c) Interferes with DNA
d) All of the above

Q40. Site of Selenium deposit in the body
a) Subcutaneous tissue
b) Hair
c) Ergot
d) Mushroom

Q41. Antidote to mercurial toxicity is
a) Butyl alcohol
b) British Anti Lewsite
c) Biotin
d) Bromocristin

Q42. Therapeutic agent for preventing carbamate toxicity is

a) 2-PAM b) AChE
c) Neostigmine d) Atropine

Q43. A cyanogenic toxic glycoside

a) Amygdalin b) Desferioxamine
c) Ferritin d) Dhurrin

Q44. A polycyclic unsaturated coumarin with bifuran &pentenone structure

a) Phenothiazine b) Aflatoxin
c) Nicotine d) Trichothecene

Q45. An estrogenic mould toxin

a) Rubratoxins b) Zearolenone
c) Ergonovine d) Aflatoxin

Q46. Maloxan, a metabolite of malathion is pharmacologically

a) Inert b) Less active
c) More active d) Unchanged

Q47. D-penicillamine acts as an antidote of copper by

a) Chemical interaction b) Competitive blockade
c) Chelation d) Dipole-dipole interaction

Q48. Iron component in cytochrome is unaffected by

a) Desferioxamine b) Dismutase
c) Magnesium ion d) Copper

Q49. Which of the following barbiturates is most rapidly metabolized in the liver

a) Pentobarbitone b) Phenobarbitone
c) Butobarbitone d) All of the above

Q50. For IV anaesthesia, best barbiturate is

a) Phenobarbitone b) Cyclobarbitone
c) Thiopentone d) Barbitone

Q51. Chlorpromazine has following effects except

a) Antitryptaminergic action b) Antihistaminergic action
c) Anticholinergic action d) Antidopaminergic action

Q52. Which of the following benzodiazepines is not changed into active metabolite?

a) Diazepam b) Flurazepam
c) Chlorazepate d) Chlordiazepoxide

Q53. Which of the following agents does not induce microsomal enzymes

a) Diazepam
b) Chloral hydrate
c) Pentobarbitone
d) Glutethiamide

Q54. Main complication of thiopentone sodium anaesthesia

a) Respiratory depression
b) Renal failure
c) Suppression of cough reflex
d) Laryngospasm

Q55. During parturition, halothane is not preferred because of

a) Fetal asphyxia caused by uterine contraction
b) Fetal respiratory depression
c) Fetal malformation
d) Inhibition of uterine contraction

Q56. Depth of anaesthesia during third stage is determined by noting characteristic changes in the following, except

a) Respiration
b) Analgesia and character of pulse
c) Spontaneous eye ball movement
d) Reflexes and muscle tone

Q57. Cardiac arrythmias are more liable to occur with the following anaesthetic, except

a) Cyclopropane
b) Thiopentone
c) Halothane
d) Trichloroethylene

Q58. Neuroleptanalgesia is

a) A combination of ether and aspirin
b) A combination of droperidol and fentanyl
c) A combination of nitrous oxide and morphine
d) A combination of diazepam and an antihistamine

Q59. Features of dissociative anaesthesia are following ,except.

a) Increase in muscle tone
b) Decrease in blood pressure
c) Respiratory stimulation
d) Slow recovery

Q60. In clinical practice, the solution of a local anaesthetic usually contains one of the following vasoconstrictor, except.

a) Epinephrine
b) Norepinephrine
c) Phenylephrine
d) Dopamine

Q61. Local anaesthetic primarily act upon

a) Sodium channels
b) Potassium channels
c) Calcium channels
d) All of the above

Q62. Antagonist of detomidine is

a) Naloxone
b) Nalorphine
c) Yohimbine
d) Atropine

Q63. Which of the following is a carbonic anhydrase inhibitor

a) Acetazolamide
b) Dichlorphenamide
c) Methazolamide
d) All of the above

Q64. Main site of action of high ceiling diuretics is

a) Proximal convoluted tubules
b) Thin ascending limb of henle
c) Thick ascending limb of henle
d) Descending limb of henle

Q65. Amino acid of angiotensin II at carboxylic end is

a) Aspirate
b) Proline
c) Tyrosine
d) Phenylaline

Q66. Atropine is generally used for pre anaestheticmedication for following purposes except

a) To potentiate hypothermia
b) To reduce bronchial secretion
c) To reduce secretion from salivary gland
d) To reduce likelihood of cardiac arrythmia

Q67. Omeprazole blocks gastric acid secretion by inhibiting

a) Na+, K+-ATPase in the parietal cell membrane
b) H+, K+-ATPase in the parietal cell membrane
c) Both of the above
d) None of the above

Q68. The binding of drugs to receptors involves

a) Ionic bond
b) Hydrogen bond
c) Van der Waals force
d) All of the above

Q69. A drug that inhibits egg production by helminths

a) Bephemium
b) Phenothiazine
c) Tetramisole
d) Ivermectin

Q70. Arsenic inactivates:

a) Na+, K+-ATPase
b) Catalase
c) Lipoic acid
d) None of the above

Q71. Opening of calcium channels is inhibited by

a) Electrical depolarization
b) Interaction with Gs proteins
c) Potassium
d) None of the above

Q72. This amino acid is not required in synthesis of Park neucleotide

a) Alanine
b) Glutamate
c) Serine
d) Lysine

Q73. Pyrental acts by this mechanism

a) Inhibiting fumerate reductase
b) Irritant
c) Depolarizing neuromuscular blocker
d) Inhibiting glucose uptake

Q74. It acts as ionophore at normal dose

a) Amikacin
b) Lasalocid
c) Benzalkonium
d) Pyrantel

Q75. It is used as both antiviral and antineoplastic drugs

a) Rifampin
b) Streptoricin
c) Bleomycin
d) Nystatin

Q76. In sheep bronchi, histamine causes

a) Constriction
b) Relaxation
c) Both depending on concentration
d) None of the above

Q77. Which of the following is a powerful inducer of platelet aggregation?

a) PGA2
b) PGI2
c) TXA2
d) LTB4

Q78. Most of the common aspirin like drugs are irreversible inhibitors of

a) Lipooxygenase
b) Monooxygenase
c) Cyclooxygenase
d) None of the above

Q79. The enzyme respsonsible for conversion of cyanide to thiocyanate in the body is:

a) Urease
b) Rhodanese
c) Acetyltransferase
d) Diaphorase

Q80. In heavy metal poisoning, dimercaprol is given as an antidote by:

a) Oral route
b) I.V. route
c) I.M. route
d) S.C. route

Q81. The most commonly used fish anaestheticis :

a) Enflurane b) Tricaine

c) Isoflurane d) Ethyl chloride

Q82. The designation of P_{450} for cytochrome involved in microsomal oxygenation comes from

a) Molecular weight b) The absorption of light

c) Atomic weight d) None of the above

Q83. Angiotensin influence urine formation by

a) Contraction of arterioles surrounding glomerulus

b) Decrease in the permeability of filtration apparatus

c) Stimulation of Na+/H+ exchange in proximal tubule

d) Decrease of Na+/H+ exchange in proximal tubule

Q84. Release of renin does not occur when there is:

a) Sodium ion depletion b) Potassium ion depletion

c) Fall in blood pressure d) Fall in blood volume

Q85. Cardiac glycosides are found in a number of plants and a few of glycosides are also present in the venom of certain

a) Snakes b) Toads

c) Mammals d) Fish

Q86. Presently the commercial heparin is prepared from mammalian

a) Lung b) Liver

c) Kidney d) Intestine

Q87. Delayed neurotoxicity is not seen with

a) Triorthocresyl phosphate b) Haloxon

c) Chlordane d) Butiphos

Q88. Suitable antidote for toxicity of zinc is

a) N.S.S. b) 5% KCl solution

c) 10% Sod. Thiosulphate solution d) 1% Sod. Carbonate solution

Q89. Xylazine produces analgesic effect acting as

a) α2 adrenergic agonist b) α 2 adrenergic antagonist

c) α1 adrenergic agonist d) α1 adrenergic antagonist

Q90. Benzodiazepines activate

a) D1 Receptor b) D2 Receptor

c) GABA Receptor d) 5-HT Receptor

Q91. Kappa opioid receptors have higher binding affinity with

a) Ketocyclazocin b) SKF 10047
c) Enkephalin d) Morphine

Q92. Na+-K+ ATPase in basolateral membrane is also called

a) Sodium-Potassium antiport b) Sodium-Potassium symport
c) Sodium-chloride symport d) Sodium-chloride antiport

Q93. Enalapril produces vasodialation by inhibiting the synthesis of

a) Angiotensin I b) Angiotensin II
c) Angiotensin d) All of the above

Q94. Organic lead is mainly stored in:

a) Hair b) Liver
c) Bone d) All of the above

Q95. In case of organophosphorus compound poisoning the immediate cause of death is due to :

a) Cardiovascular failure b) Respsiratory failure
c) Both of the above d) None of the above

Q96. One of the following is considered to be the biological magnifiers of toxic substances :

a) Bone b) Livcr
c) Fat d) Spleen

Q97. The half-life of CaNa2 EDTA in blood is:

a) 200 to 6 minutes b) 20 to 60 minutes
c) 200 to 600 minutes d) None of the above

Q98. Nitrite given in cyanide poisoning causes:

a) Relieving of CN- induced vasospasm.
b) Formation of methemoglobin
c) Both of them
d) None of them

Q99. Death from salt poisoning is due to :

a) Metabolic lesions in vital organs b) Cerebral edema
c) Both of them d) None of them

Q100. The amount of acetic acid given in the treatment of urea poisoning to a cow is:

a) 13.8 L b) 3.8 L
c) 1.38 L d) 38 L

Answer Keys

1	c	2	d	3	a	4	a	5	c	6	a	7	d
8	a	9	c	10	c	11	c	12	c	13	a	14	d
15	a	16	b	17	b	18	d	19	d	20	b	21	a
22	a	23	c	24	d	25	a	26	d	27	a	28	c
29	a	30	c	31	c	32	d	33	b	34	b	35	b
36	b	37	d	38	b	39	b	40	b	41	b	42	d
43	a	44	b	45	c	46	c	47	c	48	a	49	a
50	a	51	a	52	b	53	a	54	d	55	d	56	b
57	b	58	b	59	b	60	d	61	a	62	c	63	d
64	c	65	d	66	a	67	b	68	d	69	b	70	c
71	d	72	c	73	c	74	b	75	c	76	b	77	a
78	c	79	b	80	c	81	b	82	b	83	d	84	b
85	b	86	a	87	c	88	d	89	a	90	c	91	a
92	a	93	b	94	b	95	b	96	c	97	b	98	c
99	b	100	b										

7

Toxicity of Important Agrochemicals and Their Detoxification

Atul Prakash, Amit Shukla, Soumen Choudhury and Sakshi Tiwari

Q1. Alpha naphthyl thiourea is less toxic

a) In an empty stomach
b) With oily food
c) Full stomach
d) Not affected

Q2. A rodenticide that blocks tricarboxylic acid cycle by competitive inhibition of the enzyme aconitase

a) Fluoroacetamide
b) Pindone
c) Chloralose
d) All the above

Q3. ANTU is less toxic in an empty stomach a) In an empty stomach it will irritate the stomach wall resulting in vomiting and throwing away the poison and hence no toxicity.

a) Absorb only with fatty materials
b) Immediate absorption and detoxification
c) All the above
d) None of the above

Q4. Dimethyl dichloro vinyl phosphate (DDVP) is also called

a) Parathion
b) Dichlorvos
c) Pyrolon
d) All the above

Q5. Fumigants are used to control

a) Insects
b) Soil nematods
c) Rodents
d) All the above

Q6. Gama isomer of BHC is known as

a) Chlordane
b) Aldrin
c) Lindane
d) All the above

Q7. Infants born to mothers exposed to excessive levels of polychlorinated biphenyls are termed as

a) Pink baby
b) Blue baby
c) Cola coloured baby
d) All the above

Q8. In zinc phosphide poisoning examination of stomach content reveals

a) Odor of acetylene
b) Odor of bitter almonds
c) Odor of methane
d) None of theabove

Q9. 'Mad dog running' is a symptom of

a) Sodium fluoroacetate
b) Zinc phosphide
c) Antu
d) Warfarin

Q10. Organophosphorus inhibits

a) Cytochrome-C
b) Rhodanase
c) SGOT
d) Cholinesterase

Q11. One of the following is an organophosphorus compound

a) Carbofuran
b) HCH
c) DDT
d) Malathion

Q12. One of the following is in the decreasing order of toxicity to pyrethroids.

a) arthropods—fish--birds—mammals
b) arthropods-birds- mammals-fishes
c) mammals—birds-fishes- arthropods
d) fishes— birds—arthropods—mammals

Q13. One of the following isomers of HCH is known as Lindane

a) Alpha
b) Beta
c) Gamma
d) Delta

Q14. One of the following insecticides is most resistant to environmental degradation

a) Malathion
b) Cypermethrin
c) Phopoxur
d) DDT

Q15. Pentachlorophenol is an anti-termite compound

a) Used to preserve timbers
b) Absorbed through intact skin
c) None of the above
d) All the above

Q16. Pentachlorophenol is a

a) Fungicide
b) Molluscicide
c) All of the above
d) Herbicide

Q17. Pseudo cholinesterase is abundantly seen in

a) Blood b) Bone

c) Neuromuscular junction d) Kidney

Q18. Sodium fluoroacetate will act by

a) Inhibiting clotting of blood b) Inhibiting calcium metabolism

c) Inhibiting citric acid cycle d) None of the above

Q19. The only known naturally occurring carbamate ester

a) Carbaryl b) Eserine

c) Octachlor d) All the above

Q20. dose of a tropine in O.P. poisoning can be reduced by simultaneous administration of

a) Calcium b) 2 PAM

c) Coramine d) None of the above

Q21. The drug of choice in the treatment of parathion poisoning is

a) BAL b) 2-PAM

c) Penicillamine d) Sodium EDTA

Q22. The toxicity of zinc phosphide is due to

a) Zinc b) Phosphorus

c) Phosphine d) All the above

Q23. The rate of absorption of DDT from higher to lower is

a) Insect cuticle, mammalian gut, mammalian skin

b) Mammalian gut, insect cuticle, mammalian skin

c) Mammalian skin. Mammalian gut, insectcuticle

d) All the above

Q24. Yellow phosphorus on exposure to air

a) Emit white fumes

b) Emit fumes with a garlic odour

c) Luminous in dark

d) All the above

Q25. Yellow-green colouration of urine is seen in

a) Dinitro compounds b) Dinitro compounds

c) Phenothiazine d) none of the above

Q26. 2-4-dinitro phenol

a) Stimulate tissue respiration

b) Impair the ATP synthesis

c) Uncouple the oxidative phosphorylation

d) All theabove

Q27. Alpha naphthyl thio urea was withdrawn from the market becauseofits toxicity

a) Carcinogenicity b) Mutagenicity

c) Hepatotoxicity d) Nephrotoxicity

Q28. Among mammals .... are more susceptible to OC insecticide than other species

a) Dogs b) Cattle

c) Cats d) Pigs

Q29. Dark green colored urine isseenin .... poisoning

a) Phenol b) Cresol

c) Cyanide d) Nitrate

Q30. In organophosphorus poisoning pupilswill

a) Dilate b) Constrict

c) Exophthalmia d) None of the above

Q31. Mottling of teeth iscommonin ...... toxicity

a) Fluorine b) Chlorine

c) Nitrate d) Nitrite

Q32. Originally O.P. compounds weredevelopedas .... agents

a) Warfare b) Pesticides

c) Herbicides d) Rodenticides

Q33. Pyrethrins willacton .... membrane of the insect and killit

a) Chitin b) Mitochondrial

c) Neuronal d) None

Q34. The half-life of DDT insoilis .... years

a) 3-10 b) 1-2

c) 11-15 d) 7-10

Q35. The percentage of lindane recommended as an insecticideis ...... emulsion

a) 5% b) 10%

c) 20% d) 30%

Q36. Vitamin K is the specificantidote for

a) Organophosphates
b) Cyanide
c) Coumarin
d) Carbamates

Q37. Warfarin reduces the synthesis of ..... in liver

a) Thrombin
b) Prothrombin
c) Platelets
d) RBCs

Q38. .... is the specific antidote for organo phosphorus and carbamate insecticides

a) Pralidoxime
b) Methylene Blue
c) Thiosulphate
d) None

Q39. Rotenone is used as a

a) Bioherbicide
b) Insect hormone
c) Natural insecticide
d) Natural herbicide

Q40. This is a third-generation pesticide

a) Pheromones
b) Pathogens
c) Carbamates and organophosphates
d) Insect repellants

Q41. Amitraz is classified under

a) Organochlorines
b) Organophosphates
c) Pyrethroids
d) Formamidines

Q42. The antidote for propoxur toxicity is

a) Pralidoxime
b) Atropine sulphate
c) Diacetyl monoxime
d) Thiamine HCl

Q43. The chemical constituent commonly found in commercially available mosquito repellants is

a) Parathion
b) Allethrin
c) Amitraz
d) Bromadiolone

Q44. Which organic synthetic herbicide is often considered dangerous because it induces accumulation of nitrites in some weed species?

a) Paraquat
b) Glyphosate
c) Lindane
d) 2,4-dichlorophenoxy acetic acid

Q45. Which category of insecticidal compounds presents a problem of persistent residues in fatty tissues of animals

a) Carbamates
b) Organochlorines
c) Organophosphates
d) Pyrethrins

Q46. If acute organophosphate insecticide poisoning is suspected, what is the best initial sample to obtain from a live animal for initial diagnostic testing?

a) Serum
b) Whole blood
c) Urine
d) Stomach contents

Q47. Cholinesterase inhibitor pesticides typically cause all of the following except:

a) Salivation
b) Dyspnea
c) Blindness
d) Bradycardia

Q48. When applied to organophosphate insecticide poisoning, the term aging refers to:

a) Loss of insecticidal activity with time
b) Isomerization of the organophosphate to a more toxic chemical form
c) Hydrolysis of the cholinesterase organophosphate bond induced by Oxime drugs
d) A chemical change that increases the stability of the organophosphate Cholinesterase bond

Q49. Newer anticoagulant rodenticides, also known as second-generation anticoagulants, are important in veterinary medicine because they:

a) Have been developed to be toxic in rats but not in other classes of mammals
b) Have effects that are readily treated by synthetic vitamin k injection
c) Are more potent or longer-acting than first-generation anticoagulants, requiring prolonged therapy
d) Are more readily detected by chemical analysis than first-generation rodenticides

Q50. Methoxychlor is less toxic than

a) DDT
b) Perthane
c) Heptachlor
d) Endrin

Q51. Organochlorine insecticides protect against the acute toxicity of several organophosphorus insecticides by

a) Stimulating enzymatic detoxification of organophosphates
b) Stimulating enzymatic detoxification of organophosphates
c) Stimulating enzymatic detoxification as well as increasing noncatalytic Binding sites of organophosphates
d) None

Q52. Carbamate insecticides interact with
a) Esteratic site of acetylcholine enzyme (ache)
b) Anionic site of ache
c) Esteratic as well as anionic site of ache
d) All the above

Q53. Cypermethrin-induced toxicity is laboratory animals is also known as
a) Turner's syndrome
b) T-syndrome
c) X-disease
d) CS syndrome

Q54. The diagnostic symptom of chronic pyrethroid toxicity is
a) Muscular twitching
b) Grinding of teeth
c) Hypothermia
d) Irritation

Q55. Vitamin K is recommended in the treatment of poisoning due to
a) Sweet clover
b) Zineb
c) *Prunus* sp.
d) *Lotus* sp.

Q56. OPIDN (OP-induced delayed neuropathy) is caused due to inhibition of
a) DNA Gyrase
b) Neurotoxic esterase
c) Phosphodiesterase
d) Cholinesterase

Q57. The least toxic carbamate that is used in veterinary use is
a) Carbaryl
b) Aldicarb
c) Carbofuran
d) Ethienocarb

Q58. The specific antidote for carbamate poisoning is
a) 2-PAM
b) DAM
c) Atropine
d) All the above

Q59. Pyrethroids are natural insecticides obtained from flowers
a) Marigold
b) Sunflower
c) Rose
d) Lily

Q60. The most widely used household insecticides are
a) CHC
b) OPIs
c) Pyrethroids
d) Rodenticides

Q61. Which one is type-II pyrethroids
a) Allethrin
b) Pyrethrin
c) Permethrin
d) Cypermethrin

Q62. Which one is type-I pyrethroids
a) Deltamethrin
b) Cypermethrin
c) Fenvalerate
d) Prallethrin

Q63. The first commercially available pyrethroid used till date for repelling mosquitoes is

a) Deltamethrin b) Allethrin
c) Cypermethrin d) Fenvalerate

Q64. The species that is highly susceptible to pyrethroids toxicity is

a) Cat b) Fish
c) Dog d) Horse

Q65. The relationship between toxicity of pyrethroids and ambient temperature is

a) Inverse b) Positively related
c) Additive d) None

Q66. Pyrethroids cause CNS stimulation through inhibiting closure of

a) Voltage gated Na Channel b) Voltage gated Ca Channel
c) Voltage gated K Channel d) Voltage gated Cl Channel

Q67. Bait shyness and tolerance develops very rapidly for rodenticide

a) Warfarin b) ANTU
c) Zinc Phosphide d) Fluoroacetate

Q68. The rodenticide red squill is obtained from the plant

a) Urginea maritima b) *Chrysanthemum* spp.
c) *Physostigma* spp. d) None

Q69. The main symptom in ANTU poisoning is

a) Hypersalivation b) Convulsions
c) Fever d) Pulmonary edema

Q70. The toxic glycosidic principle present in red squill is

a) Digoxin b) Scilliroside
c) Abrin d) Nerin

Q71. Which one among these belongs to dichlorophenylethane organochlorine

a) Dicofol b) Aldrin
c) Endosulfan d) Lindane

Q72. Which one among these belongs to chlorinated cyclodiene organochlorine

a) Dalapan b) Pendimethalin
c) Atrazine d) Aldrin

Q73. 2,4,5-T belongs to

a) Dinitro compounds b) Phenoxyacetic acids
c) Triazenes d) Substituted ureas

Q74. Which one is bipyridinium compound

a) Paraquat b) Dalapon
c) Monouron d) Simazine

Q75. Which one is not a rodenticide

a) Warfarin b) Zinc phosphide
c) Fluoroacetate d) Diquat

Q76. Which one produces oestrogen-dependent tumors in lab animals

a) DDT b) Aldrin
c) Perthane d) Methoxychlor

Q77. Who discovered DDT?

a) Paul Muller b) M J B Orfila
c) Paracelsus d) Magendie

Q78. DDE excreted in

a) Exhaled air b) Urine
c) Sweat d) Faeces

Q79. Which one is used to induce hepatic microsomal enzymes to promote metabolism and excretion in organochlorine toxicity

a) Pentobarbital b) Phenobarbital
c) Secobarbitone d) None

Q80. The first organophosphate discovered is

a) TEPP b) Sevin
c) Tabun d) Paraxone

Q81. Direct acting OPIs

a) Sumithion b) Fentrothion
c) Malathion d) Fenthion

Q82. Contact OPIs

a) Parathion b) Malathion
c) Paraoxon d) All the above

Q83. Which one does not produce demyelination of the peripheral nerves and white matter of spinal cord

a) DFP b) Mipafox
c) TOCP d) TEPP

Q84. Frothy saliva is characteristics of

a) OPIs b) CHIs
c) Carbamates d) Pyrethroids

Q85. Miosis occurs in

a) OPIs
b) CHIs
c) Pyrethroids
d) Warfarin

Q86. Organophosphate insecticides induces excess of

a) Acetylcholine
b) Adrenaline
c) Dopamine
d) 5-HT

Q87. Carbamates inhibit aliesterase enzyme in addition to acetylcholinesterase enzyme in

a) Birds
b) Insects
c) Mammals
d) Amphibians

Q88. Hydrolysis of pyrethroids is done by

a) Carboxylesterases
b) Phosphodiesterases
c) AChE
d) BuChE

Q89. Burrowing behaviour seen in the toxicity of

a) Type-I pyrethroids
b) Type-II pyrethroids
c) CHIs
d) OPIs

Q90. Dinitro compounds interfere with

a) ETC
b) Dephosphorylation
c) Na-K ATPase
d) Ca channels

Q91. Drown in its own fluids is seen in

a) Zinc phosphide
b) ANTU
c) Red squill
d) Warfarin

Q92. To dilate pulmonary vessels alpha adrenoreceptor antagonists are given in treatment of

a) CHIs
b) ANTU
c) OPIs
d) Red squill

Q93. Which one accelerates synthesis of prothrombin

a) Vitamin K
b) Vitamin K epoxide
c) Vitamin K hydroquinone
d) All the above

Q94. Best example of lethal synthesis

a) Fluorocitrate
b) Warfarin
c) Atrazine
d) Sumithion

Q95. The manufacturing contaminants that make 2, 4-D extremely toxic are

a) Thiamine
b) Dioxins
c) Dinitrophenols
d) All the above

Q96. The most toxic among the herbicides are

a) Bipyridyl group
b) Dinitro compounds
c) Phenoxyacetic acids
d) Heterocyclic compounds

Q97. Urine is chrome-yellow in color and turns black upon exposure to air in

a) Bipyridyl group
b) Dinitrophenol compounds
c) Phenoxyacetic acids
d) None

Q98. Death in carbamate insecticide poisoning occurs primarily due to

a) Respiratory paralysis
b) Tachyarrythmia
c) Severe hypertension
d) Cerebral oedema

Q99. Phosphorus atom in organophosphate insecticide is

a) Monovalent
b) Divalent
c) Trivalent
d) Pentavalent

Q100. Cyclodiene insecticides stimulates neuronal excitation and convulsions by

a) Stimulation of voltage gated Na+ channels
b) Inhibition of calcium activated K+ channels
c) Activation of PLC
d) Inhibition of GABA receptors

Answer Keys

1	a	2	a	3	a	4	b	5	d	6	c	7	c
8	a	9	b	10	d	11	d	12	a	13	c	14	d
15	d	16	c	17	a	18	c	19	b	20	b	21	b
22	c	23	a	24	d	25	b	26	d	27	a	28	c
29	a	30	b	31	a	32	a	33	c	34	a	35	c
36	c	37	b	38	a	39	c	40	a	41	d	42	b
43	b	44	d	45	b	46	b	47	c	48	d	49	c
50	c	51	c	52	c	53	d	54	a	55	a	56	b
57	a	58	c	59	a	60	c	61	d	62	d	63	b
64	d	65	a	66	a	67	b	68	a	69	d	70	b
71	a	72	d	73	b	74	a	75	d	76	a	77	a
78	b	79	b	80	a	81	a	82	d	83	d	84	b
85	a	86	a	87	b	88	a	89	b	90	a	91	b
92	b	93	c	94	a	95	b	96	a	97	b	98	a
99	d	100	b										

8

Drugs Acting on Different Body Systems

Atul Prakash, Amit Shukla and Soumen Choudhury

Q1. Halothane is metabolized in to
- a) Trifluoracetic acid
- b) Inorganic chloride
- c) Bromine
- d) All the above

Q2. Psychomotor stimulants causes
- a) Excitement
- b) Relieve fatigue
- c) Increase motor activity
- d) All the above

Q3. Repeated administration of barbiturate causes
- a) Tolerance
- b) Drug dependence
- c) Toxicity
- d) All the above

Q4. One of the following drugs is a major inhalant anesthetics
- a) Halothane
- b) Diethyl ether
- c) Nitrous oxide
- d) Enflurane

Q5. Imipramine blocks the uptake of
- a) Nor epinephrine
- b) 5HT
- c) Dopamine
- d) All the above

Q6. In Oedema the choice among xanthinesis
- a) Caffeine
- b) Theophylline
- c) Theobromine
- d) All the above

Q7. Theobromine is present more in
- a) Coffee
- b) Cocoa
- c) Tea
- d) None of the above

Q8. Xanthines causes diuresis by
- a) Interfere re absorption of sodium andchlorine
- b) Specific relaxation of afferent vessels
- c) More glomeruli is put into use
- d) All the above

Q9. Psychotomimetics are
- a) Hallucinogenic
- b) Less effect of brain and spinal cord
- c) Changes the thinking pattern and mood
- d) All the above

Q10. Among the following which one is preferred inneurosurgery
- a) Halothane
- b) Enflurane
- c) Methoxyflurane
- d) Nitrous oxide

Q11. Least toxic among the following is
- a) Halothane
- b) Desflurane
- c) Methoxyflurane
- d) Sevoflurane

Q12. Brevitalis a barbiturate compound having action for
- a) Long duration
- b) Intermediary duration
- c) Short duration
- d) Ultrashort duration

Q13. The following barbiturate compounds are white in colour except
- a) Barbitone sodium
- b) Pentobarbitone sodium
- c) Pentothal sodium
- d) Diallyl barbituric acid

Q14. Barbiturates combine with other CNS depressants and
- a) Cause severedepression
- b) Accelerate the disappearance of oral anticoagulants
- c) Early metabolism of testosterone
- d) All the above

Q15. The specific antagonist of barbiturate is
- a) Bemegride
- b) 4-aminopyridine
- c) Yohimbine
- d) None of the above

Q16. The essential pre anesthetic agent in most cases
- a) Siquil
- b) Atropine
- c) Acepromazine
- d) None of the above

Q17. The neuromuscular blocking action of barbiturate can be blocked by
- a) Tubocurarine
- b) Nicotine
- c) Calcium
- d) All the above

Q18. Action of Barbiturate is by
- a) Central inhibitory transmission process mediated by GABA
- b) Suppress Glutamate induced depolarization

c) Suppress the calcium depended release of neurotransmitter
d) All the above

Q19. The duration of ultra short acting barbiturate can be prolonged with the administration of
a) 5% dextrose
b) 20% dextrose
c) 50% dextrose
d) Repeat dose of barbiturate

Q20. Guaifenesin posses
a) Anesthetic action
b) Antipyretic action
c) Expectorant action
d) All the above

Q21. One of the following is a dissociative anaestheticagent
a) Tiletamine
b) Barbiturate
c) Largactil
d) None of the above

Q22. Ketamine is a dissociative anaesthetic agent acts by
a) Antagonist at NMDA receptors
b) It stimulates sigma receptors
c) It inhibits GABA binding to CNS
d) All the above

Q23. Release of transmitters in the CNS require
a) The influx of calcium ions in to presynaptic terminals
b) The influx of calcium ions in to postsynaptic terminals
c) None of the above
d) Both 1 and 2

Q24. Following are neuropeptide transmitters of CNS
a) Endorphins
b) Substance-P
c) Angiotensin-II
d) All the above

Q25. One of the following is a pituitary peptide acts as neurotransmitters
a) Prolactin
b) GABA
c) Glycine
d) Angiotensin-II

Q26. An excitatory amino acid neurotransmitter
a) Glutamate
b) GABA
c) Glycine
d) None of the above

Q27. An inhibitory amino acid neurotransmitter
a) Aspartate
b) Glycine
c) Glutamate
d) None of the above

Q28. A hormonal neurotransmitter
a) Vasopressin
b) Glutamate
c) GABA
d) Aspartate

Q29. Cattle breed in which Idiopathic epilepsy is seen as an inherited condition

a) Jersey b) Vechur

c) Swiss brown d) None of the above

Q30. Glycine as a transmitter mainly seen in

a) Spinal cord b) Medulla

c) Cortex d) Hypothalamus

Q31. Glycine is an inhibitory neurotransmitter seen in

a) Retina b) Spinal cord

c) Lower brain stem d) All the above

Q32. On Dopaminergic neurons Substance P is having

a) An excitatory action

b) An inhibitory action

c) No action

d) Action differs on different sites

Q33. Acetyl choline is having

a) Cardiac inhibition b) Vasodilation

c) Skeletal muscle stimulation d) All the above

Q34. Acetyl choline is not used therapeutically because

a) Not available in market

b) Very short duration of action

c) No selective therapeutic response on varioustissues

d) None of the above

Q35. Acetyl choline produce vasodilatation by

a) Inhibition of the release of nor epinephrine from the sympathetic nerve terminals by activating M2 receptors

b) Interaction withM3receptors on the endothelial cells to release Nitric oxide which initiate the relaxation of vascular smooth muscles

c) Both the above

d) Direct relaxation

Q36. Alpha 2 adrenergic receptors activation in CNS causes

a) Analgesia b) Sedation

c) Skeletal muscle relaxation d) All the above

Q37. Alpha receptors are seen in

a) Blood vessels and Radial muscles of eye

b) Gastrointestinal tract

c) Splenic capsules
d) All of the above

Q38. Among catecholamines the most potent stimulator of beta1receptoris
a) Adrenaline
b) Nor adrenaline
c) Isoproternol
d) None of the above

Q39. Autonomic nervous system
a) Have an intrinsic mechanism which enable it to act automatically
b) Not absolutely essential for the action of various organs
c) Necessary for co-ordinate integrated action
d) The centers are located in hypothalamus
e) All of the above

Q40. Beta 2 receptors are located on pre-junctional neuronal membranes and postjunctional membrane of target organs like
a) Blood vessels of skeletal muscles
b) Liver Bronchi
c) Uterus, Intestine
d) All of the above

Q41. Carbachol is contra indicated in
a) Spasmodic colic
b) Intestinal obstruction
c) Pregnancy
d) All the above

Q42. Dopamine is the major neurotransmitter in the following regions
a) Basal ganglia
b) Limbic system
c) Chemoreceptor trigger zone
d) All the above

Q43. Which of the organs receive only parasympathetic division of nervous system.
a) Ciliary muscles
b) Gastric glands
c) Pancreatic glands
d) All the above

Q44. Which are choline neurotransmitters in the CNS
a) Acetyl choline
b) Butyl choline
c) Propionyl choline
d) All the above

Q45. Which drugs are depolarizing muscle relaxant
a) Tubocurarine
b) Gallamine
c) Pancuronium
d) Suxamethonium

Q46. High dose of the following drugs activates Alpha 1 receptor
a) Xylazine
b) Detomidine
c) Romifidine
d) All the above

Q47. Most of the organs have both sympathetic and para sympathetic supplyexcept

a) Adrenal medulla b) Pilomotor muscles
c) Sweat glands d) All the above

Q48. N N nicotinic receptors are present on the following areas

a) Adrenal medulla b) Spinal cord
c) Certain areas of brain d) All the above

Q49. Following drugs are synthetic para sympatholytic agents

a) Homatropine and Methyl atropine b) Methantheline
c) Dicyclomine d) All of the above

Q50. The most potent Alpha receptor stimulant

a) Adrenaline b) Nor adrenaline
c) Isoproterenol d) None of the above

Q51. All the following drugs are bronchodilators except

a) Cocaine b) Atropine
c) Histamine d) Ephedrine

Q52. A direct respiratory stimulant is

a) Epinephrine b) Ketamine
c) Ephedrine d) Doxapram

Q53. A drug used as a nasaldecongestant is

a) Isprotopium b) Terbutaline
c) Albuterol d) Phenylephrine

Q54. torant, when nebulizedan dinhaled breaks disulfide bonds with in tracheal mucosa

a) Guaifenesin b) N-acetyl cysteine
c) Potassium iodide d) Saline

Q55. A drug with non-selective bronchodilatoractionis

a) Terbutaline b) Clenbuterol
c) Albuterol d) Epinephrine

Q56. An anticholinergic drug that would cause bronchodilatation in Horse with heaves

a) Ipratropium b) Clenbuterol
c) Triamcinolone d) Atropine

Q57. An agent that most likely to increase airway resistance in a Dog with pulmonary obstruction.

a) Albuterol b) Isoproterenol
c) Propranolol d) Phenoxybenzamine

Q58. Antitussives are used in

a) Productive cough b) Dry cough
c) Both type of cough d) None of the above

Q59. All the following are mucolytic in action except

a) Sodiumacetylcysteine b) Doxapram
c) Pancreatic dornase d) Trypsin

Q60. All the followingare membrane shrinking drugs except

a) Phenylephrine b) Xylometazoline
c) Theobromine d) Oxymetazoline

Q61. A drug that is used as a cytoprotectant in peptic ulcer

a) Aspirin b) Cimetidine
c) Misoprostol d) Omeprazole

Q62. Anti muscarinics such as ……. can be used as an anti-sialagogues

a) Pilocarpine b) Atropine
c) Both d) None of them

Q63. An opiate analogue with its action limited to gut, which is recommended in diarrheais

a) Loperamide b) Butorphanol
c) Domperidone d) Meperidine

Q64. A drug which inhibits proton pump in parietal cells is

a) Famotidine b) Ranitidine
c) Omeprazole d) Diazepam

Q65. An agent which is ineffective in producing closure of oesophageal groove in calf is

a) Water b) Milk
c) 10% sodium bicarbonate d) 5% copper sulphate

Q66. Alimentary demulcentsare used

a) To mask theunpleasanttaste b) To stabilizeemulsion
c) As a suspendingagent d) All the above

Q67. Astringents usually are

a) Salts of light metal b) Organic salt
c) Salts of heavy metals d) Activated charcoal

Q68. Abomasal displacement occur mostly in

a) Middle aged dairy cow b) Heifers
c) During parturition d) None of the above

Q69. Following drugs are relatively specific to stimulate Gastric juice via H2 receptors.

a) Histamine
b) Acetyl choline
c) Pyrazole thylamine
d) None of the above

Q70. Chemoreceptor trigger zone is not involved in the emesis caused by

a) Apomorphine
b) Morphine
c) Copper sulphate
d) Ergot alkaloids

Q71. Estrogen is structurally related to

a) Bile acid
b) Cardiac glycoside
c) Vitamin- D
d) All the above

Q72. Following are androgen produced by testis other than testosterone.

a) Dihydrotestosterone
b) Androstenedione
c) Dehydroepiandrosterone
d) All the above

Q73. Gonadotropins stimulates the

a) Growth of Graffian follicles
b) Growth and development of seminiferous tubules
c) Spermatogenesis
d) All the above

Q74. In the followingspecies progesterone is not produced by placenta in latepregnancy

a) Cattle
b) Pig
c) Goat
d) All the above

Q75. In female interstitial cell stimulating hormone is involved in

a) Estrogen secretion
b) Follicle maturation
c) Ovulation
d) All the above

Q76. Oxytocin is strongly active when estrogen is in

a) Low level
b) High level
c) Normal level
d) No influence

Q77. Oxytocin is:

a) A steroid
b) Nonpeptide
c) Effective orally
d) Atropine block the action

Q78. Pregnant mare serum gonadotropins can be recommended in

a) Inactive ovary that are dormant for variousreasons
b) Suboestrum and Super ovulation for embryo transfer
c) Deficiency of sperms
d) All of the above

Q79. Synthetic estrogen belongs to the following group

a) Steroidal
b) Non- steroidal
c) Both the above
d) None of the above

Q80. The drug of choice for induction of therapeutic abortion is:

a) Oxytocin
b) Prostaglandin
c) ADH
d) All the above

Q81. An anti-mineralocorticoid used as diuretic is

a) Spironolactone
b) Aldosterone
c) Deoxycorticosterone
d) Corticosterone

Q82. A urinary acidifier

a) Ammonium sulphate
b) Sodium acid phosphate
c) Ammonium hydroxide
d) All the above

Q83. Bicarbonate can be recommended in

a) Hyperventilation
b) Uncontrolled diabetes
c) Ethacrynic acid toxicity
d) All the above

Q84. During diuretic therapy refractiveness is seen when alkalosis develops

a) For mercurials
b) For xanthenes
c) For acetazolamide
d) None of the above

Q85. Ethacrynic acid is a loop diuretic

a) It is a powerful diuretic in cats
b) It is a powerful diuretic in dogs
c) It is not a potent diuretic in rats
d) It is a mild diuretic in cattle

Q86. Ethanol is having influence on urinary system

a) Enhances ADH
b) Irritate urinary tract
c) Inhibits ADH
d) Decreases uric acid excretion

Q87. For better action of methenamine urine must be

a) Acidified
b) Alkalinized
c) Neutralised
d) Increase the volume

Q88. Furosemide is a diuretic

a) It is an anthranilicacidderivative
b) Loopdiuretic
c) Effective even in impaired renal function
d) All the above

Q89. Hexamine is a urinary antiseptic

a) Act by releasing formaldehyde in the urinary tract
b) Effective in acidic medium
c) Inhibits the action by urea splitting bacteria
d) All the above

Q90. Mersalylact as a diuretic by

a) Inhibiting the action of sod. Re absorption without interfering with na + h+ exchange
b) Interfering with na + h+ exchange
c) Inhibiting adh secretion
d) None of the above mechanism

Q91. All the following drugs predisposes the individual to digoxin toxicity , except

a) Use of loading dose
b) Hypokalemia
c) Renal disease
d) Cholestyramine

Q92. Most potent cardiac glycoside

a) Digitoxin
b) Oubain
c) Digoxin
d) Digitalin

Q93. Hypotension can be brought about by

a) Ganglion blocking drugs
b) Sympathetic blocking drugs
c) Cortical depressants
d) All the above

Q94. Ephedrine produce vasoconstriction by

a) Stimulation of sympathetic ganglia
b) Stimulation of peripheral adrenergic nerve endings
c) Direct action on blood vessels
d) Tachyphylaxis

Q95. Dobutamine is used in cardiac affection

a) It is a B2 agonist
b) It is a B1 agonist
c) It is a Beta and alpha agonist
d) None of the above

Q96. Digoxin should not be used in the following conditions

a) Ventricular fibrillation and Renal and hepatic failure
b) Circulatory shock
c) Heart block other than Congestive heart failure
d) All of the above

Q97. Digitalis cause
   a) Negative chronotropic effect
   b) Decreased automaticity
   c) A.v.node depression and Positive ionotropic effect
   d) All of the above

Q98. Acidic urine will influence the excretion of Amphetamine.
   a) It increases the excretion
   b) It decreases the excretion
   c) No change
   d) Amphetamine is not excreted in urine

Q99. Antagonist of adrenaline on heart and peripheral resistance
   a) Propranolol   b) Phenoxy benazmine
   c) Phentolamine   d) Salbutamol

Q100. An important anti angina drug is
   a) Amyl nitrate   b) Sodium nitrite
   c) Terbutaline   d) None of the above

Q101. Aglycon can be released from the cardiac glycosides by
   a) Acid hydrolysis   b) Enzymatic hydrolysis
   c) Both a and b   d) Nonc of the above

Answer Keys

1	d	2	d	3	d	4	a	5	d	6	b	7	c
8	d	9	d	10	b	11	b	12	c	13	c	14	d
15	a	16	b	17	c	18	d	19	b	20	d	21	a
22	d	23	a	24	d	25	a	26	a	27	b	28	a
29	c	30	a	31	d	32	a	33	d	34	c,b	35	c
36	d	37	d	38	c	39	e	40	d	41	d	42	d
43	d	44	d	45	d	46	d	47	d	48	d	49	d
50	a	51	c	52	d	53	d	54	b	55	d	56	a
57	c	58	b	59	b	60	c	61	c	62	b	63	a
64	c	65	a	66	d	67	c	68	d	69	c	70	c
71	d	72	d	73	d	74	d	75	d	76	b	77	b
78	d	79	c	80	a	81	a	82	b	83	b	84	a
85	c	86	c	87	a	88	d	89	d	90	a	91	d
92	b	93	d	94	b	95	b	96	d	97	d	98	a
99	a	100	a	101	c								

9

Practice Set-3

Amit Shukla, Soumen Choudhury and Sakshi Tiwari

PART-A

1. Indicate True/False for the following

1. Super infection phenomenon is common with antibiotics that are given as depot preparation. __________
2. A combination of penicillin and chlortetracycline leads to antagonism of antimicrobial activity. _________
3. Bactericidal drugs act most effectively on toxins librated by organism. _______
4. Most of the laboratory sensitivity tests are conducted at a pH of 4-5. _____
5. Streptomycin is most active at a pH 8.5 of urine _____.
6. Slow and fast acetylators of sulphonamides depend on the dose of the drug. ____
7. For treatment of urinary tract infection the doses of sulfonamides are less than that required for systemic infection.______
8. In penicillin hypersensitivity the sensitizing substance are protein conjugates of penicillin's. _______
9. Erythromycin is bacterio-static in low concentration but bactericidal in high concentration. _______
10. Sulphonamides are primarily metabolised by ethereal sulfate formation. ________
11. Amantadine competes with thymidine during synthesis of DNA for its antiviral action. _______
12. Griseofulvin is applied locally in ring worm infection because it is not absorbed from GIT. _______
13. Idoxuridne acts as antiviral agent by preventing the entry of virus into host cells. _______
14. Nitrofurautoin is useful in urinary tract infection because it is rapidly metabolized in liver which prevents effective plasma concentration. _____

15. A combination of Diamphenetide and Rafoxanide is preffered for liver fluke infection.__________

16. Calves are more prone to phenothiazine photosensitivity than adult ones.______

17. The urine of animal is brownish red following phenothiazine medication.______

18. Haloxon is the safest organophosphorus compound is cattle.________

19. Phenothiazine is not contraindicated during late stages of pregnancy.________

20. Oxyclozanide interferes with the oxidative process in parasite.________

Answer Keys

True	False	True	False	True	False
1	F	2	T	3	F
4	F	5	T	6	F
7	T	8	T	9	T
10	F	11	F	12	F
13	F	14	T	15	T
16	T	17	T	18	T
19	T	20	T		

PART-B

1. Indicate True/False for the following

1. Hormone analogues are naturally occurring substances but slightly different from hormones. ___________
2. The mechanism of action of hormones involves c AMP as second messenger ___________
3. The precise technique of estimation of hormones in the body fluids is radio immunoassay. ________
4. Thyroglobulin is produced by thyroid glands __________
5. There is no hypothalamic control for thyroid glands. _______
6. Protein bound iodine levels reflect the thyroxin secretion in the blood. ______
7. Thyroid stimulating hormones regulates iodine uptake. _______
8. Thyroxin binding plasma proteins are decreased during pregnancy. __________
9. Thyroxin binding plasma proteins are increased in nephrotic syndrome. ______
10. The half-life of thyroxin is shorter than triiodothyronine. _______
11. In goiter endemic areas iodide 1 part in 100,000 is added in the salt. ______
12. Phenyl butazone produces hypothyroidism as their side effect. ________
13. Insulin is a polypeptide hence resistant to destruction by gastric juice. ______
14. Insulin causes reduction in blood sugar level by decreasing glucose absorption from the gut. ______
15. Vasopressin is secreted from anterior Pituitary gland. _________
16. Growth hormone is secreted from Post-Pituitary gland. _______
17. Oxytocin assists in the movement of spermatozoa in the uterine cavity. _______
18. The dose of vasopressin to produce antidiuretic action is less than that required to produce pressor effect. ______
19. Synthetic corticotrophin causes less allergic reaction. _____
20. Corticotrophin is preferred over corticosteroids for long term therapy in young animals. _____

Answer Keys

True	False	True	False	True	False
1	F	2	T	3	T
4	F	5	F	6	F
7	T	8	F	9	F
10	F	11	T	12	T
13	F	14	F	15	F
16	F	17	T	18	T
19	T	20	T		

PART-C

1. Indicate True/False for the following

1. Digitoxin is excreted mainly in unchanged form. _______
2. Digitalis toxicity is enhanced by thiazide. _________
3. Single nephron glomerular filtration rate is equated with ultra filtration coefficient and difference between transcapillary pressure and mean colloidal osmotic pressure in glomerular capillaries. _________
4. Potassium enhances toxicity of cardiac glycosides. ______
5. Large doses of acetazolamide decrease gastric acid secretion. ______
6. Na+- K+- 2cl- symport inhibitors block Ca++ and Mg++ reabsorption in thick ascending limb. _______
7. Concentration of prorenin in circulation is one tenth than that of active enzyme. _____
8. Minodixil is not active hypertensive as such but become active under the influence of hepatic sulphotransferase. _______
9. In impaired renal function osmotic diuretics do not produce diuresis. ______
10. Amiloride is more potent on weight basis than Triamterene. ______
11. Hypotension is a side effect of potent diuretics. ______
12. Procaine is used as antiarrythmicagent. _______
13. Angioedema is a serious, fatal postcaptopril treated syndrome in hypertensive patient. ___________
14. Digitalis glycosides are selective inhibitors of Na+- K+ and act by binding on β subunit of Na+- K+ ATPase. ______
15. Calcium enhances the action of cardiac glycoside. ___________
16. Interaction of drugs like quinidine with digoxin may reduce volume of distribution and renal clearance rate of this glycoside. Thus loading and maintenance doses of digoxin might be reduced. _________
17. K+sparing triamterene like canrenoneantagonises aldosterone in the collecting duct. ______
18. Beta blockers are effective in terminating acute attack of angina. ______
19. Esomolol used in arrthymias has a long elimination half-life of 90 min. It produces prolonged hypertension. _______
20. A synthetic 3 hydroxy 3 methyl glutaryl coenzyme A reductase inhibitor, fluvastatin reversibly blocks cholesterol synthesis. _______

Answer Keys

True	False	True	False	True	False
1	F	2	T	3	T
4	F	5	T	6	T
7	F	8	T	9	F
10	T	11	T	12	F
13	T	14	F	15	T
16	T	17	F	18	F
19	F	20	T		

PART-D

1. Indicate True/False for the following

1. Digitoxin is excreted mainly in unchanged form. _______
2. Digitalis toxicity is enhanced by thiazide. _________
3. Single nephron glomerular filtration rate is equated with ultra filtration coefficient and difference between transcapillary pressure and mean colloidal osmotic pressure in glomerular capillaries. _________
4. Potassium enhances toxicity of cardiac glycosides. ______
5. Large doses of acetazolamide decrease gastric acid secretion. ______
6. Na+- K+- 2cl- symport inhibitors block Ca++ and Mg++ reabsorption in thick ascending limb. _______
7. Concentration of prorenin in circulation is one tenth than that of active enzyme._____
8. Minodixil is not active hypertensive as such but become active under the influence of hepatic sulphotransferase. _______
9. In impaired renal function osmotic diuretics do not produce diuresis. ______
10. Amiloride is more potent on weight basis than Triamterene. ______
11. Hypotension is a side effect of potent diuretics. ______
12. Procaine is used as antiarrythmicagent. _______
13. Angioedema is a serious, fatal postcaptopril treated syndrome in hypertensive patient. ___________
14. Digitalis glycosides are selective inhibitors of Na+- K+ and act by binding on β subunit of Na+- K+ ATPase. ______
15. Calcium enhances the action of cardiac glycoside. ___________
16. Interaction of drugs like quinidine with digoxin may reduce volume of distribution and renal clearance rate of this glycoside. Thus loading and maintenance doses of digoxin might be reduced. _________
17. K+sparing triamterene like canrenoneantagonises aldosterone in the collecting duct. ______
18. Beta blockers are effective in terminating acute attack of angina. ______
19. Esomolol used in arrthymias has a long elimination half-life of 90 min. It produces prolonged hypertension. _______
20. A synthetic 3 hydroxy 3 methyl glutaryl coenzyme A reductase inhibitor, fluvastatin reversibly blocks cholesterol synthesis. _______
21. The fenestrated endothelial cells, basement membrane and slit diaphragm are filtration barriers for negative charged macromolecules only. ________

22. The hydrostatic pressure in Bowman's Space can be equated with pressure in distal tubules. ________
23. Primary site of action for inhibitors of Na+ - Cl- symport is late distal tubule and collection duct. ________
24. Verapamil is not indicted as antiarrhythmic drug ___________
25. Hydralazine relaxes epicardial coronary artery and venous smooth muscles. _____
26. In treatment of acute aortic dissection, β adrenergic antagonist is not necessarily required with sodium nitroprusside. ______
27. The route of administration of mannitol as diuretic is oral ________
28. The increased force of heart contraction from digitalis glycoside could be a sequale of high cytosolic Ca++. _______
29. Losartan blocks angiotensin II (AT1) receptors ___________
30. Amrinone inhibits cyclic AMP hposphodiesterase to ultimately produce vasodilation. ______

Answer Keys

True	False	True	False	True	False
1	F	2	T	3	T
4	F	5	T	6	T
7	F	8	T	9	F
10	T	11	T	12	F
13	T	14	F	15	T
16	T	17	F	18	F
19	F	20	T	21	F
22	F	23	T	24	F
25	F	26	F	27	F
28	T	29	T	30	T

PART-E

1. Indicate True/False for the following

1. Super infection phenomenon is common with antibiotics that are given as depot preparation. ___________
2. A combination of penicillin and chlortetracycline leads to antagonism of antimicrobial activity. __________
3. Bactericidal drugs act most effectively on toxins librated by organism. _________
4. Most of the laboratory sensitivity tests are conducted at a pH of 4-5. __________
5. Streptomycin is most active at a pH 8.5 of urine _________.
6. Slow and fast acetylators of sulphonamides depend on the dose of the drug. _____
7. For treatment of urinary tract infection the doses of sulfonamides are less than that required for systemic infection. _______
8. In penicillin hypersensitivity the sensitizing substance are protein conjugates of penicillins. ________
9. Erythromycin is bacterio-static in low concentration but bactericidal in high concentration. ________
10. Sulphonamides are primarily metabolised by ethereal sulfate formation. _________
11. Amantadine competes with thymidine during synthesis of DNA for its antiviral action. ________
12. Griseofulvin is applied locally in ring worm infection because it is not absorbed from GIT. ________
13. Idoxuridne acts as antiviral agent by preventing the entry of virus into host cells. _______
14. Nitrofurantoin is useful in urinary tract infection because it is rapidly metabolized in liver which prevents effective plasma concentration. ________
15. A combination of Diamphenetide and Rafoxanide is preferred for liver fluke infection. _________
16. Calves are more prone to phenothiazine photosensitivity than adult ones. ______
17. The urine of animal is brownish red following phenothiazine medication. ______
18. Haloxon is the safest organophosphorus compound in cattle. ________

19. Phenothiazine is not contraindicated during late stage of pregnancy. ______
20. Oxyclozanide interferes with the oxidative process in parasite. _________

Answer Keys

True	False	True	False	True	False
1	F	2	T	3	F
4	F	5	T	6	F
7	T	8	T	9	T
10	F	11	F	12	F
13	F	14	T	15	T
16	T	17	T	18	T
19	T	20	T		

PART-F

1. Indicate True/False for the following

1. CD4+ TH2 subset of T helper cell plays a major role in the activation of Macrophages against intracellular bacteria. ____________
2. Each Clq molecule must bind to a single immunoglobulin heavy chain to be activated. ____________
3. Microbes often mutate the surface antigens that are targets of neutralizing antibodies. ____________
4. IFN-γ is the major macrophage activating cytokine. ____________
5. Immature B lymphocyte possess membrane bound IgM and IgD. _______
6. IL 4, IL 5, IL 6, IL 10 are the interleukins secreted by TH1 subset of T lymphocytes. ____________
7. Superantigens bind within the TCR antigen binding cleft. ____________
8. Both CTLs and NK cells release perforins after interacting with target cells. __________
9. Exogenous antigens are produced in altered self cells and displayed with class I MHC molecules on the cell surface. ____________
10. C5a and C3a are the anaphylotoxins produced in the complement cascade. ____________
11. TCR remains cell bound and is never secreted as BCR. ____________
12. The basic biological function of cell mediated immune system is to destroy abnormal cells. ______________
13. Natural killer a cells may also help in the production of antibodies. ________
14. Calcineurin inhibitors are one of the major immunosuppressive class of drugs used. __________
15. Monoclonal and Polyclonal antibody preparations target specifically non immune reactive cells. __________
16. Glucocorticoids lyse and induce the redistribution of lymphocytes causing lymphopenia. __________
17. Glucocorticoods have profound effect on humoral immunity. __________
18. Cyclosporine is most effective against T-cell – dependent immune mechanism underlying transplant rejection and auto immunity. __________
19. Cyclosporine increases P-glycoprotein activity. __________
20. Ca++ channel blocker affecting CYP3A interacts with the cyclosporine. __
21. The genes coding for the two chains of immunoglobulin (heavy and light chains) are located on two different chromosomes. ______________

22. Thymus-independent antigens can elicit antibody production by B cells without T cell help. ____________
23. Arthur reactions can be transferred with antiserum, whereas cytotoxic reactions are transferred with sensitized cells. ______________
24. Humoral immune response plays a major role in immunity of fungal infections than cell mediated immunity. ______________
25. Most bacterial polysaccharides are homopolysaccharide consisting of a single structural limit. ______________
26. Intracellular bacteria are sensitive to lysosomal enzymes. ___________
27. T helper cell is CD4+ whereas CTL is CD8+. ____________
28. Free circulating antibodies and those bound to antigen can initiate classical complement pathway. ____________
29. TNF-α and IL-1 result in septic shock. ____________
30. MAC alone is insufficient to kill many Gram negative bacteria. _________

Answer Keys

True	False	True	False	True	False
1	F	2	F	3	T
4	T	5	F	6	F
7	F	8	T	9	F
10	T	11	T	12	T
13	F	14	T	15	F
16	T	17	F	18	T
19	T	20	T	21	T
22	T	23	F	24	F
25	F	26	F	27	T
28	F	29	T	30	T

PART-G

1. Indicate True/False for the following

1. The biological magnifiers of toxic substances are muscles ___________
2. 10-5 to 10-7 cm is the wavelength range of Infrared rays _____________
3. The permissible level of lead in air is 0.45 mg/m^3 ______________
4. The average concentration of corbonmonooxide in the atmosphere is 0.1 ppm. ___________________
5. The major toxic effect of Benzene is on GI tract. ___________
6. WHO recommendation for limit of lead intake in food per week is 3 mg. ________
7. Nitrate concentrations in plants is enhanced by low soil temperature. ______
8. The environmental metabolism of arsenic consists of methylation-demethylation cycle. _________
9. Mercurous salt is soluble in water. _____________
10. Chlorinated hydrocarbons can be used in dust formulations. _________
11. Fat is considered as natural 'sink' for flouride. __________
12. The by product of plastic industry is pentachlorphenol. _____________
13. Cadmium is emitted into the atmosphere pre-dominatly as elemental cadmium. _______
14. Ozone is the major by product of internal comustion from vehicular traffic. _____
15. Freezing of animal fat inactivates DDT. __________
16. Dichlorvos, when sprayed on foliage is rapidly lost from the surface of leaf by oxidation. ________
17. Burning coal contaminates animal feed with cadmium. _________
18. 3 ppm is the tolerance level of chlordane in adipose tissue of food producing animals. __________
19. Who recommendation for limit of lead intake in drinking water is 1mg/1it. ______
20. Cattle can eat coal tar without any appreciable.side effects. ________
21. Anaemic persons are more susceptible to 'Co' poisoning than individuals with normal ammount of haemoglobin. __________
22. In sever 'Co' poisoning treatment includes administration of hyperbaric oxygen _________
23. Dehydrogenases are the enzymes most utilized in toxicity assays __________

24. The half life of mercury in hair is much larger than in blood __________
25. Bioaccumulation of dichlorvos occurs in different environmental compartments & organisms. ________
26. Avian and mammalian species appear to be tolerant to toxic effects of polychlorinated biphenyls. __________
27. Selenium is an important factor in metabolic defence against oxidative stress. ____
28. Flouridedoesnot interfere with calcium metabolism in developing teeth. _______
29. The active form of arsenic is pentavalent form. _______
30. Arsenic trichloride can be absorbed through skin. _________

Answer Keys

True	False	True	False	True	False
1	F	2	F	3	T
4	T	5	F	6	T
7	T	8	T	9	F
10	T	11	F	12	F
13	T	14	T	15	T
16	F	17	T	18	F
19	F	20	F	21	T
22	T	23	T	24	F
25	F	26	T	27	T
28	F	29	F	30	T

PART-H

1. Indicate True/False for the following

1. Oily solutions are suitable for intravenous injection. _____
2. Iontophoresis is the topical method of promoting absorption of drug _____
3. Drugs cross placenta primarily by active transport _____
4. The solution of local anaesthetic usually contains dopamine. _____
5. A constant fraction of drug is eliminated per unit of time for a drug following 'zero order kinetics' _____
6. Potency is directly proportional to dose. _____
7. Pentobarbitone does not induce microsomal enzymes _____
8. Fentanyl is combined with Droperidol in the ratio of 50:1 _____
9. Dissociative anaesthesia can be produced by Diazepam _____
10. Phenytoin is bound to plasma protein to a very low extent. _____

Answer Keys

True	False	True	False	True	False
1	F	2	T	3	F
4	F	5	F	6	F
7	F	8	F	9	F
10	F				

PART-I

1. Indicate True/False for the following

1. Edrophonium is used in the treatment of myasthenia gravis ____
2. Blood vessles are innervated with parasympathetic nerve fibers. ________
3. In wide angle glaucoma atropine is strictly contraindicated ____
4. Amiloride is more potent than triamterine ____
5. Carbachol is longer acting than methacholine ____
6. Tetradotoxin increases the permeability of sodium ions ____
7. Batrachotoxin decreases the permeability of sodium ions. ____
8. Organophosphates inhibits cholinesterase by phosphorylating the anionic site of the enzyme. ____
9. Danthron is more valuable purgative in horses, as its action is confined to large intestine ____
10. Aging of organophosphorus-Acetyl cholinesterase enzyme complex means acquisition of an alkyl group. ____

Answer Keys

True	False	True	False	True	False
1	F	2	F	3	T
4	T	5	F	6	F
7	F	8	F	9	T
10	F				

PART-J

1. Indicate True/False for the following

1. The elimination half – life is calculated by multiplying elimination rate constant with 0.693 ____
2. In one compartment model AUC can not be calculated ____
3. Steady state is accumulation of drug in the body in multiple dosing when input and output are unequal within dosing interval ____
4. Clearance is better index of elimination than half – life ____
5. Vd and elimination rate constant are important parameters for calculation of dosage regimen ____
6. It takes ten half – lives for elimination of 99.9% of drug obeying order kinetics ____
7. A compartment is distinguishable physiological and anatomical entity ____
8. In first order kinetics the rate of elimination is not proportional to the concentration of drug ____
9. First pass effect is that the drug is metabolised in the liver upoun absorption from GI tract before reaching systemic circulation ____
10. If pH of the body fluid and pKa of the drug are equal then 100% of the drug exists in ionized form ____
11. Mixed function oxidase comprises of reduced NADPH and molecular oxygen ____
12. For a drug following zero order kinetics, increasing the size of dose will increase the half life ____
13. Volume of distribution is expressed in relation to body weight ____
14. The higher plasma protein binding of drug leads to slowing of glomerular filtration ____
15. In case of repeated dose administration the AUC (0.00) can be calculated ____
16. F value can be calculated by the formula: AUC (iv) / AUC (non iv) ____
17. One of the characteristics of drug transportation by passive diffusion is competitive inhibition ____
18. Iontophoresis the method of application of topical agent on the skin by rubbing ____
19. For plasma protein binding of drug, the albumin should have net negative charge at a plasma pH of 7.4 ____
20. Alterations of temperature and pH of the solution affect the binding rates of the drug with plasma protein ____

21. Plasma protein binding of drug retards carrier medicated transport of drugs ____
22. The volume of distribution is apparent as it reflects a literal volume of fluid in which the drug is dissolved ____
23. Clearance is defined as volume of distribution divided by elimination rate constant ____
24. The unit of Area under the curve is ug/ml/h ____
25. The unit of clearance is ml/kg/min ____
26. First order kinetics is non linear kinetics ____
27. Mercerization is the process of decreasing the particle size ____
28. 'Flip-flop' mechanism operates when the rate of absorption is not able to control the rate of drug elimination ____
29. The conversion factor for base of natural logarithm is 2.303 ____
30. The conversion factor for logarithm to base 10 (log x) from log base e is 2.713 ____
31. In the formula for calculation of half-life there is a factor 0.693 which is an antilog of in (2) ____
32. The relationship of half- life, volume of distribution and clearance is :

    t ½ = 0.693 x Vd x Cl ____
33. The relationship of half- life with mean residence time (MRT) is;

    t ½ = 0.693/MRT ____
34. The relationship of half- life with mean absorption time (MAT) is;

    t ½ = 0.693/MAT ____
35. In non compartmental analysis of plasma concentration time profile, the formula for calculating body clearance is: Cl¬B = Dose x AUC ____
36. In non compartment analysis of plasma concentration time profile, the formula for calculating volume of central compartment (Vc) is: Dose/C° p ____
37. A drug following zero order kinetics is expressed in the form of linear kinetics ____
38. The volume of distribution of a drug in young animals is less than the adults ____
39. Vd ss is dependent on the elimination processes ____
40. The plasma concentration time profile of a drug given by i.v. route of administration is always described by two compartment model are located in definite anatomical and physiological space ____

Answer Keys

True	False	True	False	True	False
1	F	2	F	3	F
4	T	5	T	6	T
7	F	8	F	9	T
10	F	11	T	12	T
13	T	14	T	15	F
16	F	17	F	18	F
19	T	20	T	21	F
22	F	23	F	24	F
25	T	26	F	27	T
28	F	29	F	30	F
31	T	32	F	33	F
34	T	35	F	36	T
37	F	38	F	39	F
40	F				

PART-K (Pharmacokinetics and Clinical Pharmacology)

1. Indicate True/False for the following

1. Most of the laboratory tests involving antimicrobial activity is conducted at a pH of 4-5.______
2. Steady state is accumulation of drug in the body in multiple dosing when input and output are unequal within dosing interval.________
3. It takes ten half-lives for elimination of 99.9% of drug obeying first order kinetics.________
4. A compartment is distinguishable physiological and anatomical entity ________
5. First pass effect is the metabolism of drug in the liver upon absorption from GI tract before reaching systemic circulation____________
6. If pH of the body fluid and pKa of the drug are equal then 100% of the drug exists in ionized from________
7. For a drug following zero order kinetics, increasing the size of dose will increase the half-life________
8. Super infection phenomenon is common with antibiotics that are given as depot preparation ________
9. F value can be calculated by the formula: AUC (iv)/AUC(non iv)________
10. For plasma protein binding of drug, the albumin should have net negative charge at a plasma pH of 7.4__________
11. Alterations of temperature and pH of the solution affect the binding rates of the drug with plasma protein________
12. Plasma protein binding of drug retards carrier mediated transport of drugs_____
13. The volume of distribution is apparent as it reflects a literal volume of fluid in which the drug is dissolved________
14. Clearance is defined as volume of distribution divided by elimination rate constant________
15. The unit of Area under the curve is µg/ml/h________
16. First order kinetics is linear kinetics________
17. 'Flip-flop' mechanism operates when the rate of absorption is not able to control the rate of drug elimination. ________
18. The relationship of half-life, volume of distribution and clearance is:
    t1/2 = 0.693 x Vd x Cl __________
19. The relationship of half-life with mean residence time (MRT) is:
    t1/2 = 0.693 /MRT __________

20. Bactericidal drugs act most effectively on toxins librated by organism. ________

21. Streptomycin is most active at a pH 8.5 of urine__________.

22. Slow and fast acetylators of sulphonamides depend on the dose of the drug.______

23. For treatment of urinary tract infection the doses of sulfonamides are less than that required for systemic infection.________

24. Sulphonamides are primarily metabolised by ethereal sulfate formation. ________

25. Amantadine competes with thymidine during synthesis of DNA for its antiviral action. _________

26. Griseofulvin is applied locally in ring worm infection because it is not absorbed from GIT._________

27. Idoxuridne acts as antiviral agent by preventing the entry of virus into host cells.________

28. Nitrofurantoin is useful in urinary tract infection because it is rapidly metabolized in liver which prevents effective plasma concentration._________

29. In penicillin hypersensitivity the sensitizing substance are protein conjugates of penicillin._________

30. A combination of penicillin and chlortetracycline leads to antagonism of antimicrobial activity.___________

Answer Keys

True	False	True	False	True	False
1	F	2	F	3	T
4	F	5	T	6	F
7	T	8	F	9	F
10	T	11	T	12	F
13	F	14	F	15	F
16	T	17	F	18	F
19	F	20	F	21	T
22	F	23	T	24	F
25	F	26	F	27	F
28	T	29	T	30	T

10

Veterinary Therapeutics and Drug Discovery

Soumen Choudhury, Amit Shukla and Atul Prakash

Q1. Which of the following is semisynthetic product:

a) Chlortetracycline b) Chloramphenicol
c) Tetracycline d) Oxytetracycline

Q2. Natural source of amphotericin B is:

a) Streptomyces aureofaciencs b) Streptomyces nodosus
c) Streptomyces erythreus d) Streptomyces venezuelse

Q3. Following is /are mitotic inhibitor

a) Paclitaxel b) Griseofulvin
c) Vinblastine d) All of the above

Q4. Vancomicin has the following unwanted effects:

a) Pseudomembranous colitis b) Hepatotoxicity
c) "Red neck" syndrome, phlebitis d) All of the above

Q5. Lincozamides have the following unwanted effect:

a) Nephrotoxicity b) Cancerogenity
c) Pseudomembranous colitis d) Irritation of respiratory organs

Q6. Which of the following drugs is used for systemic and deep mycotic infections treatment:

a) Co-trimoxazol b) Griseofulvin
c) Amphotericin B d) Nitrofungin

Q7. Which of the following agents is an antibiotic and a antineoplastic agent?

a) Gentamicin b) Amphotericin B
c) Hygromycin B d) Bleomycin

Q8. Which of the following is a source for one of the antineoplastic drugs?

a) Atropa beladone b) Digitalis lanata
c) Vinca rosea d) Strychnos nuxvomica

Q9. Hetacilillin in the body is converted to which of the following agents?

a) Ampicillin
b) Carbenicillin
c) Penicillin G
d) Penicillin V

Q10. Which of the following antibiotics has got good antifungal action?

a) Bacitracin
b) Polymyxin
c) Spectinomycin
d) Griseofulvin

Q11. Basic structure of cephalosporins is consists of:

a) Thiazolidine ring
b) Two of the above
c) Beta lactam ring
d) Dihydrothiazine ring

Q12. Tetracyclines tend to form complexes with

a) Magnesium ions
b) Calcium ions
c) Aluminium ions
d) All of the above

Q13. All of the following anthelmintics are anti-cestodal agents except:

a) Praziquantel
b) Niclosamide
c) Ivermectin
d) Bithionol

Q14. An ideal anthelmintic should have qualities like

a) Efficacy, wide therapeutic index, ease of administration and under limit residues
b) Broad spectrum of activity, safety, single dose efficacy, long duration of action
c) Specific and supportive action like diarrhea, quick excretion and long residual effect
d) Efficacy, safety, ease of administration, quick action

Q15. Which one of the following is an nalogue of guanosine

a) Acyclovir
b) Cyterabine
c) Zidovidin
d) Idoxuridine

Q16. Which one of the following acts by blocking the assembly in viral replication

a) Ribavirin
b) Rimantadine
c) Foscarnet
d) Viderabine

Q17. Nitrofurantoin is most effective as urinary antiseptic when pH of the urinary tract is:

a) Neutral
b) 5.5 or less
c) Over 9.5
d) Between 6 and 7

Q18. Which of the following is not produced by moulds?

a) Chloramphenicol
b) Penicillins
c) Chlortetracycline
d) Fluoroquinolones

Q19. Which of the following aminoglycosides is employed for local action in the GIT?

a) Gentamicin
b) Neomycin
c) Tobramycin
d) Kanamycin

Q20. Following antibiotic antagonize with ADH

a) Penicillin
b) Doxycycline
c) Minocycline
d) Both of the above

Q21. Gray baby syndrome is associated with

a) Penicillin
b) Chloramphenicol
c) Tetracycline
d) Macrolides

Q22. Most of the antibiotics are derived from

a) Moulds
b) Virus
c) Bacteria
d) Yeast

Q23. Which of the following is not recommended in horse

a) Fluroquinolone
b) Chloramphenicol
c) Amphotericin B
d) Penicillin

Q24. Following are the prodrug EXCEPT

a) Nitobimine
b) Thiophanate
c) Levodopa
d) Ampicillin

Q25. The transfer of resistance genes between bacteria occurs by the mechanism(s):

a) Conjugation
b) Transformation
c) Transduction
d) All of the above

Q26. Who of the following scientists discovered natural penicillins?

a) Alexander Flemming
b) Gerhard Domagk
c) Sir Waksman
d) Sir Henry Dale

Q27. Endecticide action of ivermectin is due to its

a) Action on glutamate gated Cl- channel
b) Disruption of reproductive function
c) Both of the above
d) None of the above

Q28. AMES test is standard in vitro test for

a) Mutagenicity
b) Carcinogenicity
c) Teratogenicity
d) All of the above

Q29. Which of the following drugs is used for dermatomycosis treatment:

a) Nystatin
b) Griseofulvin
c) Amphotericin B
d) Vancomycin

Q30. All of the following antifungal drugs are antibiotics, EXCEPT:

a) Amphotericin B
b) Nystatin
c) Miconazol
d) Griseofulvin

Q31. Azoles have an antifungal effect because of:

a) Inhibition of cell wall synthesis
b) Inhibition of fungal protein synthesis
c) Reduction of ergosterol synthesis
d) Inhibition of DNA synthesis

Q32. Amfotericin B has the following unwanted effects:

a) Psychosis
b) Renal impairment, anemia
c) Hypertension, cardiac arrhythmia
d) Bone marrow toxicity

Q33. Which of the following drugs alters permeability of Candida cell membranes:

a) Amphotericin B
b) Ketoconazole
c) Nystatin
d) Terbinafine

Q34. Characteristics of Amfotericin B are following, EXCEPT:

a) Used for systemic mycosis treatment
b) Poor absorption from the gastro-intestinal tract
c) Does not demonstrate nephrotoxicity
d) Influences the permeability of fungus cell membrane

Q35. Following is/are excreted through bile

a) Penicillin
b) Doxycycline
c) Minocycline
d) Both the above

Q36. Following inhibit polymerization of microtubule EXCEPT

a) Benzimidazole
b) Griseofulvin
c) Vincristine
d) Colistin

Q37. Which one is the drug of choice for fasciolosis

a) Oxyfenbendazole
b) Thiophenate
c) Triclabendazole
d) None of the above

Q38. Which of the following affects the permeability if the cell membrane:

a) Cycloserine
b) Bacitracin
c) Polymixin B
d) Imidazole

Q39. Synergistic action of colistin is observed with

a) Oleandomycin | b) Lincomycin
c) Erythromycin | d) Clendamycin

Q40. Anaplasmosis can be treated with

a) Oxytetracycline | b) Amoxicillin
c) Dihydrostreptomycin | d) Oleandomycin

Q41. Which of the following is used in the treatment of topical fungal infection

a) Benzoic acid | b) Salicylic acid
c) Both of the above | d) None of the above

Q42. Which of the following is used for studying cell current/ion channel current

a) Organ bath | b) Patch clamp
c) Confocal microscopy | d) RT-PCR

Q43. Drug development consist of following phases

a) 3 phases | b) 6 phases
c) 4 phases | d) 5 phases

Q44. Which one of the following is ideal for bioassay of histamine

a) Chicken crop | b) G. pig ileum
c) Frog rectus abdominis | d) Rat uterus

Q45. FIC <1 indicates

a) Synergistic effect | b) Antagonistic effect
c) Additive effect | d) None of the above

Q46. The term 'antibiotic' was coined by

a) Vuillemine | b) Waksman
c) Domagk | d) Paul Earlich

Q47. Nucleic acid metabolism of microorganism is affected by

a) Clotrimazole | b) Rifampin
c) Vidarabine | c) Cycloserine

Q48. Which of the following is most effective against Gm –ve bacteria

a) Cefuroxime | b) Cephalothin
c) Cephalaxin | d) Cefotaxime

Q49. Use of fluoroquinolones is not recommended in young and growing due to the following:

a) Idiosyncratic aplastic anaemia | b) Hepatotoxicity
c) Renal toxicity | d) Arthropathic lesions

Q50. Which of the following agents is an antibiotic having anthelmintics action?

a) Gentamicin b) Amphotericin B

c) Hygromycin B d) Bleomycin

Answer Keys

1	c	2	b	3	d	4	c	5	c	6	c	7	d
8	c	9	a	10	d	11	b	12	d	13	d	14	a
15	a	16	b	17	b	18	d	19	b	20	b	21	b
22	a	23	a	24	d	25	d	26	a	27	c	28	a
29	b	30	c	31	c	32	b	33	c	34	c	35	d
36	d	37	c	38	c	39	c	40	a	41	c	42	b
43	c	44	b	45	a	46	b	47	c	48	d	49	a
50	c												

11

Practice Set-4

Soumen Choudhury, Amit Shukla and Atul Prakash

PART-A

Q1. Elimination of drug is studied under?

a) Pharmacokinetics
b) Pharmacometrics
c) Toxicodynamics
d) Toxicoelimination

Q2. Serpentine receptor is characterized by

a) 7 Transmembrane loop
b) Petals of lilly like structure
c) Zinc fingers
d) JAK STAT

Q3. Status epilepsy is treated best with ________ lorazepm/diazepam

a) Per rectal
b) Intravenous
c) Intramuscular
d) Oral

Q4. ________ classified poisons for the first time into plants, animals, minerals and animal poisons.

a) Theophrastus
b) Orfila
c) Maimon
d) Dioscorides

Q5. Nicotinic receptors are

a) GPCR
b) Enzyme linked Receptors
c) Nuclear receptors
d) Inotropic Receptors

Q6. Zinc fingers are characteristic feature of

a) Estrogen receptor
b) Histamine receptors
c) Both of the above
d) Insulin

Q7. Autophosphorylation is a needed step in the cascade of

a) Steroid Receptors
b) Nuclear receptor
c) Both of the above
d) Tyrosine kinase linked receptors

Q8. Followings are the examples of alkaloids except"

a) Morphine b) Nicotine

c) Digoxin d) Atropine

Q9. Branch of pharmacology deals with the use of genetic information to guide the therapy is known as

a) Pharmacokinetics b) Pharmacognosy

c) Pharmacogenetics d) Pharmacogenomics

Q10. Agonist should possess

a) Affinity but no efficacy b) Only Efficacy

c) Potency but no affinity d) Affinity and efficacy

Q11. Graded dose response curve follows

a) All or None Law b) Law of Mass Action

c) Both of the above d) Ficks Law

Q12. First in patient study is conducted under

a) Phase III b) Phase I

c) Phase II d) Phase IV

Q13. Sulphation is not a common metabolic reaction of phase II in ___________ species.

a) Canine b) Bovine

c) Swine d) Feline

Q14. Ethosuximide produces its action by targeting

a) N type calcium channel b) T type calcium channel

c) P/Q type calcium channel d) L type calcium channel

Q15. Study of quantitative measurement of drug action is done in

a) Pharmacodynamics b) Pharmacoepidemiology

c) Pharmacometrics d) Posology

Q16. Which of the following is a prodrug?

a) Aspirin b) Enalapril

c) Losartan d) Captopril

Answer Keys

1	a	2	a	3	a	4	a	5	d	6	a	7	d
8	c	9	d	10	d	11	b	12	c	13	c	14	b
15	c	16	b										

Part-B

Match the following agents/drugs with their natural sources

Drug/agent	**Natural source**
1. Avermectins	A. *Micromonospora purpurea*
2. Hygromycin B	B. *Penicillium notatum*
3. Chloramphenicol	C. *Streptomyces hygroscopicus*
4. Amphotericin B	D. *Streptomyces erythreus*
5. Streptomycin	E. *Bacillus subtilis*
6. Erythromycin	F. *Streptomyces avermitilis*
7. Bacitracin	G. *Penicillium griseofulvium*
8. Penicillin G	H. *Streptomyces griseus*
9. Gentamicin	I. *Streptomyces nodosus*
10. Griseofulvin	J. *Streptomyces venezualae*

Part-B (Match the following)

i) F	ii) C
iii) J	iv) I
v) H	vi) D
vii) E	viii) B
ix) A	x) G

Part-C

Fill in the blanks with appropriate word (s)

i) .... is recognized as the Father of Chemotherapy.

ii) Nitroxinyl is highly effective against mature .... in sheep and cattle.

iii) Anthelmintic Levamisole is an ....

iv) .... is an platinum containing anticancer agent

v) Chloramphenicol is a lipid soluble compound, which lacks both the acidic and basic groups but contains ....

vi) Size of nanoparticles for drug delivery ranges from .... to ....

vii) .... is the anthelmintic which possesses exceptionally high activity against immature satges of flukes.

viii) Sulfonamides exert the antibacterial action by inhibiting the enzyme ....

ix) The primary site of action of avermectins in nematodes is ....

x) Imidocarb is recommended for the treatment of ....

Part-C (Fill in the blanks)

i) Paul Ehrlich

ii) Trematodes

iii) Immunomodulator

iv) Cisplatin

v) Nitrobenzene moiety

vi) 1 to 100 nm

vii) Diamfenetide

viii) DHF synthtase

ix) Synapse between inhibitory interneurones and excitatory motor neurones

x) Babesiosis